# The Beauty of Everyday Mathematics

Norbert Herrmann

# The Beauty of
# Everyday Mathematics

**Copernicus Books**

*An Imprint of Springer Science+Business Media*

Dr.Dr.h.c. Norbert Herrmann
Universität Hannover
Institut für Angewandte
Mathematik
Welfengarten 1
30167 Hannover
Germany
Dr.Dr.Herrmann@googlemail.com

Translated from German by Martina Lohmann-Hinner, mlh.communications

Published in the United States by Copernicus Books,
an imprint of Springer Science+Business Media.

Copernicus Books
Springer Science+Business Media
233 Spring Street
New York, NY 10013
www.springer.com

Library of Congress Control Number: 2011940689

Mathematics Subject Classification (2010): 00A09

Translation from the 3rd German edition *Mathematik ist überall* by Norbert Herrmann, with kind permission of Oldenbourg Verlag Germany. Copyright © by Oldenbourg Verlag Germany. All rights reserved.

Printed on acid-free paper

ISBN 978-3-642-22103-3        e-ISBN 978-3-642-22104-0
DOI 10.1007/978-3-642-22104-0

# Preface

Once upon a time, there was a group of representatives from the State of Utah in the United States of America around the year 1875. One of them was James A. Garfield. During a break, they were sitting in the congressional cafeteria. To pass the time, one of them, namely Mr. Garfield, suggested that they take a look at the Pythagorean Theorem. Even though this famous theorem had already been studied and proven 2000 years ago, he wanted to come up with a new proof. Together with his colleagues, he worked for a little while, and discovered the following construction:

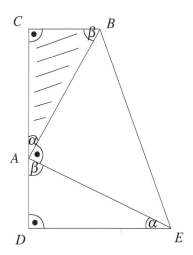

**Fig. 0.1** Sketch proving the Pythagorean Theorem.

Here, we ave the crosshatched right triangle $\triangle ABC$. We sketch this triangle once more below it, though this time turned slightly so that side $\overline{AD}$ lies exactly on the extension of side $\overline{AC}$. The connecting line $\overline{EB}$ completes the figure, turning it into a trapezoid because the bottom side is parallel to the top side thanks to the right angles. The two triangles meet, with their angles $\alpha$ and $\beta$, at $A$. Because the triangles are right triangles, the two angles add up to $90°$, from which we conclude immediately that the remaining angle at $A$ is also a right angle. After all, three angles equal $180°$ when added together.

Now, only the little task remains of comparing the area of the trapezoid (central line times height, where the central line equals (base line + top line)/2 with the sum of the areas of the three right triangles:

$$\frac{a+b}{2} \cdot (b+a) = 2 \cdot \frac{a \cdot b}{2} + \frac{c^2}{2}.$$

The simple solution of this equation provides the formula of Mr. Pythagoras:

$$a^2 + b^2 = c^2.$$

The sum of the areas of the squares on the two legs equals the area of the square on the hypotenuse.

Mr. Garfield submitted this proof for publication. And, sure enough, the proof was actually published in the New England Journal of Education. The mere fact that there had been some representatives who had occupied their spare time during a break with mathematics would have been worth mentioning.

But now comes the most extraordinary aspect. The spokesman of these math fans was the James A. Garfield who, a little later, became the President of the United States.

You just have to savor the moment. A long, long time ago, there was actually once a president of the United States who published a new proof of Pythagoras, in the nineteenth century. He not only could recite this famous theorem, but also understood it completely and even proved it.

We wouldn't dare claim that many politicians today probably consider the Pythagorean Theorem to be a new collection of bed linen. But what is so remarkable is the fact that representatives whiled away their spare time with mathematical problems. Today, any mathematician who openly proclaims his or her profession is immediately confronted with the merry message that their listener has always been bad at math.

Mr. Garfield was only President for less than a year, because he was shot with a pistol by a crazy person in Washington's train station. He died soon after the attack. Is this maybe a reason why today's presidents, kings, chancellors, etc. avoid mathematics?

I truly hope that this little book will make a small contribution towards conveying the beauty of mathematics to everyone.

I would like to thank specifically my editor, Mr. Clemens Heine. His enthusiastic response to the idea of writing this book was very helpful. Many thanks to the Assistent Editor Mathematics, Mrs. Agnes Herrmann, for her cooperation during the preparation of this edition.

Last but not least, I would like to thank my wife, who cleaned my desk at home out of desperation while I was spreading chaos elsewhere in the house.

Meissen, November 2011                                          *Norbert Herrmann*

# Contents

Contents

# Chapter 1
# The Soda Can Problem

## 1.1 Introduction

I'd just like to clarify one thing first: This author doesn't like cans, ir-respective of their content. The energy consumption during their pro-duction and recycling is just too high. But our problem can be solved much more easily with a can because of its great symmetry. A math-ematician sees a can as a cylinder. A bottle has this awkward neck, which, if it were to be described precisely, might cause us a headache. That's why we'll focus first on a can. Later on, we'll see how we can transfer some of these comments to a bottle.

## 1.2 The Problem

A group of young people has decided to picnic in the great outdoors and is now sitting on the grass somewhere, battling flies and, well, with this specific can of soda. Because this stupid container doesn't want to stay upright on the grass; instead, it just wants to tip over and spill its delicious contents so that the ants can enjoy it. This is exactly the kind of task which makes physicists and mathematicians roll up their sleeves together.

The physicist gets to speak first:

First of all, we idealize the can as a completely even cylinder, and ig-nore the tab on the top and the indentation on the bottom. We also dis-

N. Herrmann, *The Beauty of Everyday Mathematics*,
DOI 10.1007/978-3-642-22104-0_1, © Springer-Verlag Berlin Heidelberg 2012

regard the fact that the color is not distributed evenly on the surface as a result of the labeling. And we graciously ignore the tiny irregularities in the can's shape. Just consider it to be an ideal cylindrical can.

Now, if the can is still completely full, its center of gravity is located exactly in the middle, which is actually due to its symmetry; that's pretty clear, right?

Careful, now: what if the can is completely empty? Where is the center of gravity now? Right, exactly in the middle again.

But if we take a few sips from the full can, then the content (unfortunately) decreases and the center of gravity sinks to the bottom.

And since, at the end of te process, the center of gravity is once again in the middle, it's only logical that it can't continue to just keep on sinking. Instead, it turns around at some point and moves up towards the middle again.

Conclusion: there is, in other words, a state in which the center of gravity reaches its lowest point. This state has to be determined. Because that's the point where the can assumes its most stable position. That's when we can put it on the grass with the least amount of risk.

The question, therefore, is: at which level of the fluid is the center of gravity at its lowest point? Or, put another way:

---

**The Soda Can Problem**

How much soda do you have to drink so that the center of gravity reaches its lowest point?

---

## 1.3 Determining the Center of Gravity

Now, we have to briefly turn to physics.

- $M$ indicates the mass of the full can,
- $m$ the mass of the empty can,
- $H$ the total height of the can, and
- $h$ the height of the residual fluid inside the can.

Physics teaches us that a body's center of gravity is determined as follows:

$$S = \frac{\int \rho \cdot r \, dV}{\int \rho \, dV},\qquad(1.1)$$

where the integrals each need to be calculated over the entire volume of the body. Here, $\rho$ is the density of the body, i.e., the mass per unit volume:

$$\rho = \frac{m}{V}.$$

Now, it's essential to keep in mind that we're looking at a can that has a completely symmetrical body with regard to its longitudinal axis because this way, we're able to focus on just this longitudinal axis; in other words, we consider our volume integral as an ordinary integral in the longitudinal direction. We're also assuming that the fluid inside the can is completely homogeneous, and that the can is also homogeneous.

There are ten three fixed points of reference. First, the lower point of the can, which we'll call 0 because it marks the beginning of the system of coordinates. Second, the point at which the fluid currently stands is defined as $h$. Third, the upper end point of the can is called $H$. This results in

$$\begin{aligned} S &= \frac{\int \rho \cdot r \, dV}{\int \rho \, dV} \\ &= \frac{\int_0^h \rho_v \cdot x \, dx + \int_h^H \rho_l \cdot x \, dx}{\int_0^h \rho_v \, dx + \int_h^H \rho_l \, dx}. \end{aligned}$$

A word on the density. In the range 0 to $h$, it's that of the soda, including the can's material; in the range $h$ to $H$, it is just the can's material. We can move either $\rho_v$ or $\rho_l$ in front of the integral, since it is a constant. Next, we relate the density to the entire can and get

$$S = \frac{\rho_v \cdot \int_0^h x\,dx + \rho_l \cdot \int_h^H x\,dx}{\rho_v \cdot \int_0^h 1\,dx + \rho_l \cdot \int_h^H 1\,dx}$$

$$= \frac{\frac{M}{V} \cdot \int_0^h x\,dx + \frac{m}{V} \cdot \int_h^H x\,dx}{\frac{M}{V} \cdot \int_0^h 1\,dx + \frac{m}{V} \cdot \int_h^H 1\,dx}.$$

We can now multiply the numerator and the denominator by $V$ and get

$$S = \frac{\int_0^h M \cdot x\,dx + \int_h^H m \cdot x\,dx}{\int_0^h M\,dx + \int_h^H m\,dx}.$$

A little reminder of school will then hopefully confirm the following result:

$$\int x\,dx = x^2/2 + c_1, \quad \int 1\,dx = x + c_2.$$

Because of the limits of the integral, the constants $c_1$ and $c_2$ can be omitted. We thus get the overall formula for determining the can's center of gravity when the can has been emptied to the level $h$:

$$S = \frac{1}{2} \cdot \frac{Mh^2 + m \cdot (H^2 - h^2)}{Mh + m \cdot (H - h)}. \tag{1.2}$$

Note, though, that the can's bottom and top have to be treated separately; we can't just include them in the mass.

We're now able to determine the center of gravity for every partially emptied can.

We're now going to prove this by determining the center of gravity of the full can according to the formula. Our heuristic preliminary consideration actually told us that the center of gravity is located right in the middle, i.e., at $h = H/2$. So, let's see if that's true.

A full can means that $h = H$, that is, $H - h = 0$. Therefore, the second term of both the numerator and the denominator of our formula (1.2) can be omitted. This is what remains:

$$S = \frac{1}{2} \cdot \frac{Mh^2}{Mh} = \frac{1}{2} \cdot h = \frac{1}{2} \cdot H.$$

In fact, the center of gravity is actually located right in the middle.

So, let's now consider the second case, of the empty can, i.e., $h = 0$. Again, not much remains of the formula (1.2):

$$S = \frac{1}{2} \cdot \frac{mH^2}{mH} = \frac{1}{2} \cdot H.$$

Such a little test is not without value; it makes you feel as if you're not completely off with the formula.

Unfortunately, the pretty formula (1.2) doesn't give us any clues as to how much you have to drink so that the center of gravity reaches its lowest level. There are two different methods by which we can actually find this out. The first method comes from a nice little thought found in Vogel and Gerthsen's excellent book on physics [9]. The second approach is based on a slightly more monotonous analysis by setting the value of the first derivative to zero.

## 1.4  The Lowest Position of the Center of Gravity

### *Determination by Thought Experiment*

We would normally consider that a body's center of gravity can be identified by trying to balance the body on a needle. Of course, that's impossible if a fluid is involved. Therefore, we consider this fluid to be solid, for example, frozen. Then we're able to just put the can on its side and balance its center of gravity. Thanks to symmetry, a knife edge will do.

Now, we just need an idea. When we introduced this task, we didn't fully utilize one particular aspect: if the center of gravity first sinks and then rises again during drinking while the fluid level (unfortunately) continues to drop, then the center of gravity must be at the same height as the fluid level at some point in time.

Stop! That sounds like an interesting point: the center of gravity is located at the level of the fluid. In fact, that's actually the critical point. Because we're asserting:

If the center of gravity is located at the fluid level, then it will be at its lowest position.

So, in order to prove this, let's imagine that the center of gravity is located at the fluid level and that the fluid is now frozen. Then we're able to put the can on its side and balance it on a knife edge. The knife edge is located exactly below the fluid level. Now, let's assume that the part of the can which is full is on the right-hand side and that the can's empty part is on the left-hand side of the knife edge.

So, if we now imagine that some fluid is added (we'll not call it spitting into the can, but that's exactly what it is), the left side will get a bit heavier; that's why we have to move the knife edge to the left, which causes the center of gravity to move up. Well, that was pretty clear.

But if we now imagine that we're taking some (frozen) fluid out of the can – in other words, we're drinking something – then the part on the right-hand side becomes lighter, which means that the can tips towards the left side again while the center of gravity, once again, moves up.

This furnishes proof that the center of gravity couldn't move any lower – which is a really interesting thought.

Thus, we simply have to put the position of the center-of-gravity $S$ into our center of gravity formula (1.2) at the same level as the fluid level $h$, i.e., $S = h$, which now permits us to calculate $h$ from the emerging equation:

$$h = S = \frac{1}{2} \cdot \frac{m \cdot (H^2 - h^2) + M \cdot h^2}{m \cdot (H - h) + M \cdot h}. \tag{1.3}$$

Now, here's a little bit of algebra. We multiply by the denominator, and then can delete a couple of things, thus getting a quadratic equation in $h$:

$$2m(Hh - h^2) + 2Mh^2 = mH^2 - mh^2 + Mh^2,$$
$$2mHh - 2mh^2 + 2Mh^2 = mH^2 - mh^2 + Mh^2,$$
$$2mHh + Mh^2 = mh^2 + mH^2,$$
$$(M - m)h^2 + 2mHh - mH^2 = 0,$$
$$h^2 + \frac{2mH}{M - m}h - \frac{mH^2}{M - m} = 0.$$

We solve this equation using the standard formula solving quadratic equations:

$$
\begin{aligned}
h_{1,2} &= -\frac{mH}{M - m} \pm \sqrt{\left(\frac{mH}{M - m}\right)^2 + \frac{mH^2(M - m)}{(M - m)^2}} \\
&= \frac{-mH \pm \sqrt{m^2H^2 + mMH^2 - m^2H^2}}{M - m} \\
&= \frac{-mH \pm \sqrt{mMH^2}}{M - m} \\
&= \frac{(-m \pm \sqrt{mM})H}{M - m} \\
&= \frac{mH}{M - m}\left(-1 \pm \sqrt{\frac{M}{m}}\right).
\end{aligned}
$$

Together with the $-1$, the negative sign in front of the square root would generate an overall negative result, which would be meaningless from a physical point of view. What's the purpose of a negative fluid level?

That's why we select only the positive sign, and we get the result we've been looking for with the help of the third binomial formula:

$$h = \frac{mH}{M - m}\left(\sqrt{\frac{M}{m}} - 1\right) = \frac{\sqrt{m}}{\sqrt{M} + \sqrt{m}} \cdot H. \qquad (1.4)$$

## Determination Through Analytical Considerations

This is also done quite easily. We're considering the center of gravity as a function of the fluid level's height $h$:

$$S = S(h) = \frac{1}{2} \cdot \frac{m \cdot (H^2 - h^2) + M \cdot h^2}{m \cdot (H - h) + M \cdot h}. \tag{1.5}$$

According to the rules of analysis, we have to differentiate this function with respect to $h$ and set the derivative zo 0. This is an essential prerequisite for an extremum. Next, we have to show that the solution will actually provide the expected minimum by calculating the second derivative, demonstrating that it is positive at this point. But we would rather leave that to the real math freaks.

So, let's begin! We apply the quotient rule:

$$\left( \frac{f(x)}{g(x)} \right)' = \frac{f'(x) \cdot g(x) - f(x) \cdot g'(x)}{(g(x))^2}.$$

If this equals 0, then we only have to set the numerator equal to 0. This results in the following:

$$-2m^2 hH + 2m^2 h^2 - 2mMh^2 + 2mMhH - 2mMh^2 + 2M^2 h^2$$
$$- mMH^2 + mMh^2 - M^2 h^2 + m^2 H^2 - m^2 h^2 + mMh^2 = 0$$

Now, we summarize a couple of things and arrange them according to the powers of $h$:

$$(m - M)^2 h^2 + 2mH(M - m)h + (m - M)mH^2 = 0,$$

$$(m - M)h^2 - 2mHh + mH^2 = 0,$$

$$h^2 - \frac{2mH}{m - M}h + \frac{mH^2}{m - M} = 0.$$

That's a quadratic equation, which we're going to solve by completing the square:

$$
\begin{aligned}
h_{1,2} &= \frac{mH}{m-M} \pm \sqrt{\left(\frac{mH}{m-M}\right)^2 - \frac{mH^2}{m-M}} \\
&= \frac{mH}{m-M} \pm \sqrt{\frac{m^2H^2 - mH^2 + mMH^2}{(m-M)^2}} \\
&= \frac{mH}{m-M} \pm \frac{H}{m-M}\sqrt{mM} \\
&= \frac{mH}{m-M} \pm \frac{mH}{m-M}\sqrt{\frac{M}{m}} \\
&= \frac{mH}{m-M} \cdot \left(1 \pm \sqrt{\frac{M}{m}}\right) \\
&= \frac{mH}{M-m} \left(\pm\sqrt{\frac{M}{m}} - 1\right).
\end{aligned}
$$

As already mentioned above, we now recognize once again that only the positive sign in front of the square root makes sense from a physical point of view. A negative level is useless in practice. And so, we get the final result,

$$
h = \frac{mH}{M-m}\left(\sqrt{\frac{M}{m}} - 1\right) = \frac{\sqrt{m}}{\sqrt{M} + \sqrt{m}} \cdot H, \tag{1.6}
$$

which corresponds to our formula (1.4). Isn't it great how well math and physics complement each other!

## 1.5 Drinking in Two Mouthfuls

Now, the strategy for our picnic is absolutely clear.

First, we need the measurements of the can, i.e., its total height, its empty weight, and its total weight. But stop, doesn't the formula (1.4) actually require the masses? Don't panic! In the two fractions, we're going to multiply both the numerator and the denominator by the acceleration due to the earth's gravity, $g = 9.81\text{m}/\text{sec}^2$, which doesn't really change anything (we've essentially multiplied by 1), and instead of the mass, we're going to put the weight everywhere:

$$h = \frac{G_{\text{empty}} \cdot H}{G_{\text{full}} - G_{\text{empty}}} \cdot \left( \sqrt{\frac{G_{\text{full}}}{G_{\text{empty}}}} - 1 \right) = \frac{\sqrt{G_{\text{empty}}}}{\sqrt{G_{\text{full}}} + \sqrt{G_{\text{empty}}}} \cdot H.$$

$$(1.7)$$

By the way, here is an interesting insight. Instead of the acceleration due to gravity on the earth, we could have used any other nonzero number – even the acceleration due to gravity on the moon, for example. This means that our result would also pass muster on the moon. So if, dear reader, you plan to have a picnic on the moon, where the ground is also very uneven, then our formula would be equally helpful. (Did you catch the double meaning of this statement?)

Now, though, we have to undertake the following steps:

1. At home, we weigh a full can $G_{\text{full}}$ and weigh an empty can to obtain $G_{\text{empty}}$, and we get $M$ and $m$.

2. We measure the can's height $H$.

3. Next, you'll have to do some math to determine the level $h$ of the lowest point of the center of gravity. Perhaps you can mark this position on the can at home.

4. During the picnic, the challenge will then be **to empty the can in two mouthfuls**. The first mouthful has to go no farther than the mark.

5. Then you can put the can down on the grass with little or no risk.

6. When taking the second mouthful, you have to empty the can.

## 1.6 Center of Gravity of an Ordinary Can

Let's consider an ordinary half-liter can (i.e., 16.907 fl. oz.). We've determined its total mass as being $M = 537\ g$. Since the content is approx. $L/2 = 500$ g (some sodas consist mostly of water), the empty can has mass $m = M - 500$ g $= 37$ g.

We now insert this into our formula (1.4) and get

$$h = \frac{mH}{M-m}\left(\sqrt{\frac{M}{m}} - 1\right)$$
$$= \frac{37 \cdot 16}{537 - 37}\left(\sqrt{\frac{537}{37}} - 1\right)$$
$$= 3.33\ \text{cm}.$$

If we have a smaller can containing 350 ml (i.e., 11.835 fl. oz.), we get the following values:

total mass $M = 377$ g

empty mass $m = 27$ g

total height $H = 11.5$ cm

$$h = \frac{mH}{M-m}\left(\sqrt{\frac{M}{m}} - 1\right)$$
$$= \frac{27 \cdot 11.5}{577 - 27}\left(\sqrt{\frac{377}{27}} - 1\right)$$
$$= 2.43\ \text{cm}.$$

So, first you have to take a really big mouthful from our cans before putting them on the grass. Well, this actually seems to come in handy if you're thirsty.

And maybe we could convince the beverage industry to indicate the lowest position of the center of gravity on cans right from the start.

## 1.7 Final Remarks

A word about other containers. Of course, our result that the center of gravity assumes its lowest position if it is located at the fluid level also applies to packages of other fluids in the same way. And so, our result is correct as well with regard to juice and milk cartons, i.e. those containers. They're just as symmetrical when it comes to their vertical central axis. Yes, it's even true for all kinds of bottles; that's actually what makes this idea so beautiful. And yes, the bottle might even be lopsided and crooked. Just the basic dimensions need to be modified; in this case, specifically, the total height $H$. Bottlenecks become narrower towards the top. Thus, one needs to find an average value if we imagine that the bottle ends cylindrically at the top. Maybe it would suffice if we were to estimate roughly how much we need to subtract from the height so as to get an adequate approximation.

But let's not go over the top! Remember, this is just a game. So, let's say:

Cheers!

# Chapter 2
# The Mirror Problem

## 2.1 Introduction

> Mirror, mirror on the wall,
> Who is the fairest one of all?

asks the jealous queen in the fairy tale "Snow White" and gets an answer she doesn't like.

Since time immemorial, looking into a mirror has been very popular. An amusing question which keeps popping up at regular intervals in magazines is the following:

How large does a mirror have to be so that a person can see themselves completely in it?

Here are some typical answers:

1. It depends on how far I stand from the mirror.
2. The mirror has to be as tall as I am.

These seem to make sense; but, unfortunately, both answers are wrong.

## 2.2 The Mirror Problem for Individuals

Let's look first at a person who wants to buy a mirror.

N. Herrmann, *The Beauty of Everyday Mathematics*,
DOI 10.1007/978-3-642-22104-0_2, © Springer-Verlag Berlin Heidelberg 2012

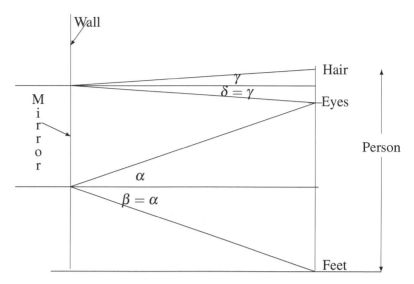

**Fig. 2.1** A person standing in front of a mirror has only to look down halfway to their feet and look up halfway to their hair in order to see themselves completely; that's why the mirror – if attached correctly – has to be only half as tall as the person.

The image in Figure 2.1 will help explain the situation.

According to the law of reflection, the angle of incidence $\alpha$ equals the angle of reflection $\beta$ and, likewise, the angle of incidence $\gamma$ equals the angle of reflection $\delta$. Now, if I look into a mirror hanging vertically on the wall directly opposite me; and I want to see my feet, then I have to lower my eyes only halfway between my eyes and the floor. My line of sight hits the mirror at the angle $\alpha$. The angle of reflection $\beta$ has the same size and automatically directs my line of sight towards my feet.

The same thing happens if I want to look at my hair. My line of sight needs to only be lifted halfway between my eyes and my hair, allowing me to see that my gray hair doesn't get any darker in the mirror either.

Thus, we can summarize this as follows:

1. Seeing yourself completely in a mirror doesn't depend on the distance you are from the mirror.

2. The mirror needs to be exactly half as tall as you are.

Some care needs to be exercised when hanging up the mirror. As mathematicians, we can summarize this in an algorithm.

### Algorithm for Attaching a mirror for an Individual

1. The mirror has to hang on a vertical wall.

2. The mirror has to be half as tall as I am.

3. The mirror has to be hung in such a way that I can just see my hair at the upper edge of the mirror. The upper edge of the mirror thus has to be mounted at a height equal to

my eye level + half the distance to my eyes − half the distance to my hair

## 2.3 The Mirror Problem for Groups

If several persons within a group, for example, a family or a bowling club, all want to look at themselves in the same mirror from top to toe, then we have to plan a little bit more carefully.

If you take the tallest person in the group, then you might be able to align the mirror at its upper edge with this person. The bottom edge is determined by half the height of this person. But if the mirror is actually attached this way, then a shorter person won't be able to see their feet. So if we don't want to attach the mirror to a movable device so that it can be pulled up and down like a hanging lamp, then we have to give the sales clerk some more money to buy a larger mirror.

The tallest person determines where the mirror is mounted at the top.

The length downwards from this point is defined by the shortest person who is actually the person with the lowest eye level. At the bottom, the mirror has to reach the halfway point of this person's eye level. This results in the total length of the mirror being as follows (figure 2.2:

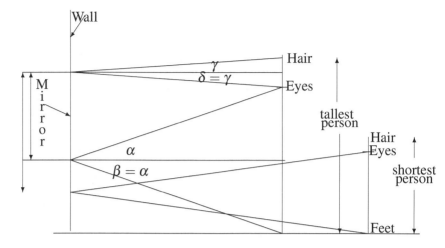

**Fig. 2.2** Here, we've placed the shortest person in the group next to the tallest person. We can see that this person has to look down beyond the bottom edge of the mirror so that they can see themselves completely. Thus, the mirror has to be taller. The arrows on the left indicate the length of the mirror for an individual and, further left, the length of the mirror for a group.

### Total Length of Group's Mirror

eye level + half the distance to the hair to eye level of the tallest person − halfway to the eye level of the smallest person.

### Algorithm for Attaching a Mirror for a Group

1. The mirror has to hang on a vertical wall.
2. The mirror has to be attached in such a way that the tallest member of the group is able to just see their own hair in the mirror.
3. Thus, the upper edge of the mirror has to be mounted at a height equal to

eye level + half the distance to the eyes − half the distance to the hair of the tallest person

Well, this was just a rehearsal, which has hopefully triggered our interest in the following, more difficult task.

## 2.4 The Problem

The question we're now going to ask is a more puzzling one:

---

### The Mirror Problem

If I shake my right hand, my mirror image shakes its left hand. But if I shake my head, my mirror image doesn't shake its feet. So why does a mirror exchange right and left, but not top and bottom?

---

We haven't found anyone yet who didn't find this question interesting. But what is the answer? To find it, we have to examine the mirror a bit more precisely; actually, mathematically.

Those of my dear readers who are not familiar with vectors in a plane might want to skip the following section. We'll summarize the result in Section 2.6. That's where those readers can log on again, because the result is understandable even to people who are mathematical amateurs.

## 2.5 The Mirror Problem Expressed Mathematically

We now enter the plane to look at vectors.

So that we don't have to carry too much of a burden with us, which, in turn, could cause us to lose sight of the essentials, we'll choose a situation where the mirror's axis is the $y$ axis. The diagram in Figure 2.3 shows that a random point $(x,y)$, which we identify with its position vector $\mathbf{a} = (x,y)$, is depicted in the mirror at a point $\mathbf{a}^* = (-x,y)$.

Such a representation has very simple properties. If we take a multiple of the vector $\mathbf{a}$, we can form the image of the vector $\mathbf{a}$ first, followed by multiplication of the image vector, which we can express mathematically as follows if we abbreviate "reflection" with the symbol $RE$:

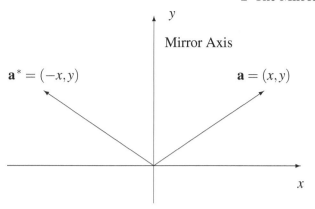

**Fig. 2.3** Reflection at the $y$ Axis

$$RE(k \cdot \mathbf{a}) = k \cdot RE(\mathbf{a}).$$

We can play the same little game when we add two vectors together. We can obtain the image of the sum if we first form the images of the two vectors separately and then add the images together, which is expressed mathematically as follows:

$$RE(\mathbf{a} + \mathbf{b}) = RE(\mathbf{a}) + RE(\mathbf{b}).$$

Such representations are called "linear."

Now, math teaches us that such a representation can best be described by a matrix. A matrix is a square field into which we enter numbers. Since we're looking at vectors in the plane, we take a $2 \times 2$ field. This field describes the representation. In order to characterize it as a mathematical rule, we enclose it in parentheses. This is done as follows:

$$\begin{pmatrix} 1 & 2 \\ 3 & 4 \end{pmatrix}.$$

In order to extract the mapping rule, we look at the two unit vectors $\mathbf{e}_1 = (1, 0)$ and $\mathbf{e}_2 = (0, 1)$ and image them:

$$\mathbf{e}_1 = (1, 0) \to (-1, 0), \qquad \mathbf{e}_2 = (0, 1) \to (0, 1).$$

We've written this down in a relaxed manner. But take a look again at the drawing in Figure 2.3. The first unit vector $\mathbf{e}_1 = (1,0)$ points exactly to the right from the origin. If it's mirrored in the $y$ axis, then its image vector points exactly to the left from the origin. That's the vector $-\mathbf{e}_1 = -(1,0) = (-1,0)$. The second unit vector, $\mathbf{e}_2 = (0,1)$ points vertically upward. It is located directly on the $y$ axis. That's why its reflection stays where it is.

Now comes the rule for the calculation scheme; that is, the matrix of the representation.

We enter the image vectors into a $(2 \times 2)$ matrix as columns:

$$A = \begin{pmatrix} -1 & 0 \\ 0 & 1 \end{pmatrix}. \tag{2.1}$$

Thus, the representation is written in the form of a matrix as

$$\mathbf{x} \to \mathbf{x}^* = \begin{pmatrix} -1 & 0 \\ 0 & 1 \end{pmatrix} \cdot \mathbf{x}.$$

Now, we have to introduce another term, which is, in principle, quite difficult to explain. Since we want to linger only in this plane, though, we've only got to consider $2 \times 2$ matrices. Our new term is in fact quite simple.

**Definition 2.1** *By the determinant of a $2 \times 2$ matrix*

$$\begin{pmatrix} a & b \\ c & d \end{pmatrix},$$

*we mean the expression*

$$\det(A) = a \cdot d - b \cdot c.$$

That's just a number that's a little bit hard to calculate. Let's use an example:

$$A = \begin{pmatrix} 2 & 4 \\ -3 & 1 \end{pmatrix} \quad \Rightarrow \quad \det(A) = 2 \cdot 1 - 4 \cdot (-3) = 14.$$

The next example shows that the determinant of a matrix can become negative:

$$B = \begin{pmatrix} 2 & 4 \\ 3 & 1 \end{pmatrix} \quad \Rightarrow \quad \det(A) = 2 \cdot 1 - 4 \cdot 3 = -10.$$

The sign of the determinant is actually the distinctive feature that we are interested in, because mathematicians have found the following:

**Theorem 2.1** *Linear representations whose corresponding matrices have a positive determinant retain orientations whereas linear representations whose matrices have a negative determinant reverse orientations.*

Well, it's slowly becoming clearer. It's all about orientation.

A brief comment for purists:

All of this shouldn't anger serious mathematicians. We have just given some hints here and there; those who wish to get more information are advised to look in specialist textbooks on linear algebra.

## 2.6 Results of Analysis of the Mirror Problem

First, we note that there are significant differences between the pairs "right and left" and "top and bottom."

The terms "right" and "left" have a significance which depends on the respective person. What is "right" for me is "left" for my counterpart, and vice versa. It is my personal orientation that determines where right and left are.

In contrast, the terms "top" and "bottom" are the same for everyone, at least for those who are in our immediate vicinity, for example all Americans. We all look up towards the clouds when someone talks to us from above.

And here's an important insight:

For our reflection matrix $A$ in equation (2.1), we get

$$\det (A) = (-1) \cdot 1 - 0 \cdot 0 = -1.$$

So the determinant is negative! And this means:

**The orientation is reversed when mirrored.**

The best way to grasp this concept is by looking at a circle and its mirror image. When we move a pen along a circular line in a clockwise direction from the top, our mirror image moves the pen in a counterclockwise direction. If I were to show my mirror image a clock face, then the clock would always run in the wrong direction, regardless of where I was standing or where the mirror was located. That's what happens with orientation.

The mirror doesn't actually swap right and left; instead, it reverses the orientation. Orientation is thus associated directly with persons. My orientation tells me where to find the right side. The mirror reverses this image, which explains why my mirror image sees this quite differently and calls it left.

"Top" and "bottom," however, are terms of an entirely different nature. "Top" and "bottom" are the same for everyone in our vicinity. They are objective terms, which are not associated with our personal orientation. These terms are, thus, not swapped by a mirror. In fact, this applies to the terms "east" and "west" as well. If I point towards the east, my mirror image also points towards the east in the mirror realm. And if I ask my mirror image, when I am standing in Berlin, to point towards the Eiffel Tower, then my mirror image will point in the mirror in exactly the same direction as I am pointing: towards Paris and not towards Moscow.

# Chapter 3
# The Leg Problem

## 3.1 Introduction

Isn't it great to be finally out strolling through the city and looking at all the happy people during the first days of spring after a long, hard winter? As their spirits soar, people start to unburden themselves of their bulky winter clothing. In fact, the clothes people wear are often as light as their spirits. You might even speculate about what's a safe distance at which admire the legs of the person walking ahead of you. Experience has shown that this can be a moral issue because the best distance is not 0 mm. That's much too close, because then the visual angle would be 0°. On the other hand, it's reassuring to know that an infinite distance is also not a solution to the problem, because from such a distance, the visual angle tends to be zero as well. The optimal distance is somewhere in between. Mathematics can provide some valuable assistance with this sensitive topic.

## 3.2 The Problem

So, the problem to be solved is as follows:

N. Herrmann, *The Beauty of Everyday Mathematics*,
DOI 10.1007/978-3-642-22104-0_3, © Springer-Verlag Berlin Heidelberg 2012

---

**The Leg Problem**

What is the distance at which one should walk behind another person so as to have the best possible visual angle of that person's legs?

---

## 3.3 The Physical Model

Let's simplify the situation to create a physical model.

- The street is absolutely even, so that we can use it as an $x$ axis.
- The person to be observed is simply a straight line. All moralists should applaud here.
- We're going to consider only the part of the line running from the ground to the hemline of the skirt or shorts; this is also assumed to be a straight line (more applause?). No other details need be considered.
- We'll also ignore the fact that we move up and down as we walk. Our eye level, therefore, always is assumed to be at the same level. The same is true for the person ahead of us.

We can also forget the course of time and limit ourselves to a momentary snapshot. We depict this situation in the Figure 3.1.

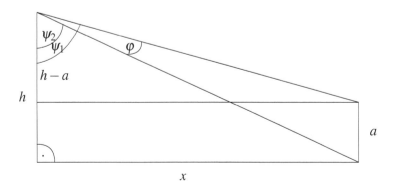

**Fig. 3.1** The leg problem, abstracted of all bodily realities.

The descriptions can be summarized as follows:

1. $a$  is the leg height to be considered.
2. $h$  is my eye level.
3. $\psi_1$  is the angle to the uppermost visible part of the leg.
4. $\psi_2$  is the angle to the lowermost visible part of the leg.
5. $\varphi$ is my visual angle, which has to be optimized.
6. $x$ is the distance we seek.

## 3.4  Analytical Solution

To derive a mathematical model, we need to recall some formulas from trigonometry.

We choose to access the angles via the tangent function. We have

$$\tan \psi_1 = \frac{x}{h-a}, \qquad \tan \psi_2 = \frac{x}{h}.$$

For angle $\varphi$ in the sketch,

$$\varphi = \psi_1 - \psi_2 = \arctan \frac{x}{h-a} - \arctan \frac{x}{h}.$$

This is a formula in which we take the eye level $h$ and the leg height $a$ as given. Only the distance $x$, which we seek, is unknown. From calculus, we know that an extremum can occur only when the first derivative vanishes. We thus calculate the first derivative and set it to zero:

$$
\begin{aligned}
\varphi' &= \frac{1}{1 + \left(\dfrac{x}{h-a}\right)^2} \cdot \frac{1}{h-a} - \frac{1}{1 + \left(\dfrac{x}{h}\right)^2} \cdot \frac{1}{h} \\
&= \frac{h-a}{(h-a)^2 + x^2} - \frac{h}{h^2 + x^2} \\
&= 0.
\end{aligned}
$$

A simple conversion leads us to the following result:

$$x_0 = \sqrt{h(h-a)}. \tag{3.1}$$

However, we still have to show mathematically that this really is a maximum. The second derivative has to be negative for this distance. The second derivative is

$$\varphi'' = -\frac{a}{(2h-a)^2\sqrt{h(h-a)}} < 0,$$

because all terms in this fraction are positive and there is a minus sign in front.

## 3.5 Graphical Solution

After this short excursion into calculus, we can now even be happier, because we can also find a geometric solution.

Let's take a look at the sketch in Figure 3.2, into which we have inserted a circle, namely a circle drawn through the three points "my eyes,""lowermost visible part of the leg,"and "uppermost visible part of the leg."A circle can be drawn through any three points which are not in a straight line. Its center in this case is directly on the perpendicular bisector of the distance $a$, which we've added as a dashed line.

Let's assume a fixed distance $a$ and a circle large enough to include this distance as a chord. A certain angle is formed by this distance when it is observed from the opposite side of the circle. If we now expand the circle a little bit, this angle will become smaller. And if we reduce the size of the circle, then the angle will increase.

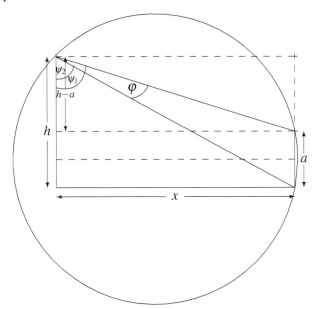

**Fig. 3.2** Our leg problem with an added circle that goes through the three points "my eyes,""lowermost visible part of the leg,"and "uppermost visible part of the leg."

We can demonstrate this with the sketches in Figure 3.3.

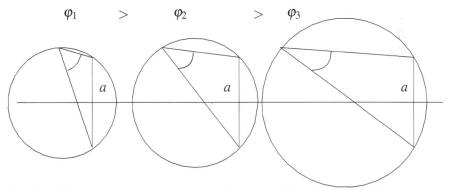

**Fig. 3.3** The visual angle depends on the size of the circle. The bigger the circle, the smaller the angle formed by the chord $a$.

Our objective therefore needs to be to find the smallest possible circle which passes through the three points mentioned above. Since our eye level is fixed – we drew a dotted line in the sketch to represent

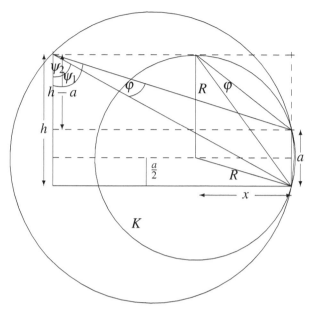

**Fig. 3.4** Graphical solution of the leg problem. The smallest possible circle, $K$, is shown here.

our eye level – the smallest possible circle is the circle that just touches this line. The line is thus the tangent to the circle. We've therefore added this circle, $K$, to the sketch in Figure 3.4.

Now we just need to calculate distance $x$ for this circle. Pythagoras is going to help us do so:

$$x = \sqrt{R^2 - \left(\frac{a}{2}\right)^2} = \sqrt{\left(h - \frac{a}{2}\right)^2 - \left(\frac{a}{2}\right)^2} \qquad (3.2)$$

$$= \sqrt{h^2 - ah + \left(\frac{a}{2}\right)^2 - \left(\frac{a}{2}\right)^2}$$

$$= \sqrt{h(h-a)}. \qquad (3.3)$$

This agrees with our analytical result (3.1).

## 3.6 Application and Comments

For our application of the above argument, let's take a look at a young man in shorts. Let the distance from the hem of his shorts to the ground be $a = 60$ cm. Let my eye level be $h = 170$ cm. The perfect distance is then

$$x_0 = \sqrt{170 \cdot (170 - 60)} = \sqrt{18\,700} = 136.75.$$

I thus have to walk behind him at a distance of about 1.40 m.

A comment on the history of this problem: It is said that Franz Rellich assigned it to his students because they had found his calculus lectures too abstract.

This problem can be found in many different disguises. One could, for example, ask how far one has to stand from a church tower in order to get the best view of the clock on the tower. Mathematically the same task, but very boring.

## 3.7 A Mnemonic Device for $\pi$

We've had so much to do with circles that we'd like to add this little treat.

Do you have difficulty remembering $\pi$? 3.14, and then what? We'll show you a method by which you can always remember four decimal places of $\pi$. Do you want to see how? You just have to say "J,"as in "jog your memory,"loud and clear. Take a look at the picture in Figure 3.5.

We start with "J" for "jog your memory."We count clockwise to see how many letters are left in each block. Then, we add a decimal point at the right location, and get

$$\pi \approx 3.1416.$$

Isn't that ingenious? At this point, I'd like to thank my friend Rein Luus [6] from the University of Toronto for proposing this neat mnemonic device.

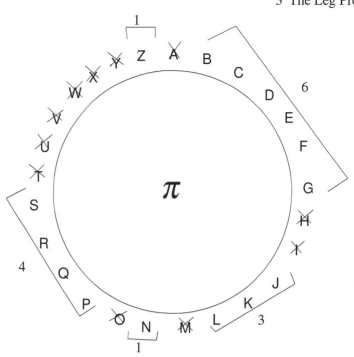

**Fig. 3.5** $\pi$ has something to do with circles. We therefore write all capital letters of the alphabet in a circle and then cross out all those letters which are symmetrical about their central vertical axis. A number of letters remain which we have divided into blocks.

## 3.8  Comments on the Number $\pi$

The number $\pi$ is determined geometrically by the ratio of the circumference of a circle to its diameter.

1. In ancient Babylonia, the approximations $\pi \approx 3$ and also $\pi \approx 25/8 = 3.125$ were used.

2. The Egyptians knew the approximation $\pi \approx (16/9)^2 = \underline{3.16}049$. (We have underlined the correct decimal digits.).

3. The Chinese found the approximation $\pi \approx 355/113 = \underline{3.14159292}$.

4. Archimedes[1] defined upper and lower limits $3\frac{10}{71} < \pi < 3\frac{10}{70}$, that is, $\underline{3.14085} < \pi < \underline{3.14286}$.

---

[1] Archimedes of Syracuse (287 B.C. to 212 B.C.).

5. Vieta[2] calculated $\pi$ to nine decimal places.

6. Ludolf van Ceulen[3] introduced the symbol $\pi$. That's why this number is sometimes also called the Ludolphine number.

7. Lindemann[4], who was born in Hannover, proved in 1882 that $\pi$ is transcendental, which is why the ancient challenge of squaring the circle, i.e. the construction of a square with the same area as a given circle with only compasses and a ruler, is impossible to solve.

---

[2] Francois Vieta (1540–1603).

[3] Ludolf van Ceulen (1540–1610).

[4] Ferdinand Lindemann (1852–1939).

# Chapter 4
# The Sketch Problem

## 4.1 Introduction

In this chapter, we'd like you to participate in a very special experiment. We want to show you why mathematicians have to be so strict when they formulate their statements.

You're all familiar with the following false conclusion:

*When it rains, the road is wet (except for tunnels or under trees, etc.).*
*But when the road is wet, it doesn't necessarily have to be because of rain.*

Maybe a sprinkler system was put in the wrong place, or maybe I was running around with my watering can.

No, another false conclusion will be identified in this chapter. Namely, the following false conclusion:

### But you can actually see it!

Oh, how many things we see are actually false! One could list hundreds of examples, but here we're going to draw a very nasty little diagram with which we can trick you.

N. Herrmann, *The Beauty of Everyday Mathematics*,
DOI 10.1007/978-3-642-22104-0_4, © Springer-Verlag Berlin Heidelberg 2012

## 4.2 The Problem

We want to solve the following problem:

---

**The Sketch Problem**

We assert that

$$90° = 100°$$

and then ask:

Where is the mistake in our reasoning?

---

But first, we'd like to remind you of a case from intermediate level. (Or were you playing "battleships" with your neighbor?) It's about the congruence of triangles. One of the four congruence theorems states the following:

**Theorem 4.1** *If the three sides of two triangles coincide, then the triangles are congruent, i.e. they are identical in all parts (same angles, same height, etc.).*

It might be possible that we can put the triangles on top of each other, and that they're then identical to one another. But it's also possible that you first have to create the mirror image of one of the two triangles before they're identical. Both possibilities are expressed by the word "congruent."

## 4.3 The "Proof"

Using the term "congruence," we'll now demonstrate that $90° = 100°$. As evidence, we'll take a look at the sketch in Figure 4.1.

In the sketch, we've marked a downward angle of $90°$ at $A$ and a downward angle of approximately $100°$ at $B$. We draw two sides of the same length at the right and left. They are, of course, not parallel. The straight line connecting $C$ to $D$ is then not parallel to the side $\overline{AB}$.

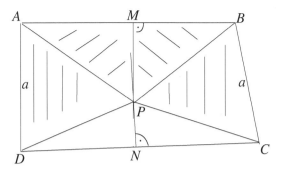

**Fig. 4.1** Sketch used as evidence that $90° = 100°$.

On the sides $\overline{AB}$ and $\overline{CD}$, we draw the mid-perpendiculars or per-pendicular bisectors, of these sides. They intersect; we call the point of intersection $P$. From $P$, we draw the connecting straight lines to $A$, $B$, $C$, and $D$.

So much for our sketch. We can note the following:

1. The following applies to the two triangles with slanted hatching:

   a. $\overline{AM} = \overline{MB}$, because the perpendicular bisector was drawn at $M$.

   b. $\overline{AP} = \overline{PB}$, because $P$ is on the perpendicular bisector of $\overline{AB}$.

   c. $\overline{PM}$ is the same in both triangles.

   Therefore all three sides are identical, and all angles match as well. In other words, the angle at $A$ equals the angle at $B$:

$$\angle PAM = \angle PBM.$$

2. The two vertically hatched triangles are also congruent since:

   a. $\overline{AP} = \overline{PB}$, because the perpendicular bisector was drawn at $M$.

   b. $\overline{DP} = \overline{PC}$, because $P$ is on the perpendicular bisector of $CD$.

   c. $\overline{AD} = \overline{BC}$, by construction.

Therefore all angles are identical here as well. The angle at *A* equals the angle at *B*:

$$\angle DPA = \angle CPB.$$

Altogether, we've proved that the total angle at *A*, which is 90°, equals the total angle at *B*, which we had made 100°. Therefore

$$90° = 100°.$$

This is, of course, nonsense. But where is our mistake? Please be assured that the congruence theorem is correct. Maybe you should now open your old textbooks again and see that the theorem is actually correct. No, we've cheated somewhere else.

## 4.4 The First Clue

We should actually let you stew a bit at this point, because it would be great if you remembered what you're learning here.

But maybe you could sit down and attempt to copy our sketch. If we take just a quick glance at it, we can see that the right angles are not quite as "right" as they might be. Maybe the point of intersection is not even inside the figure?

But wait a minute! We can achieve that as well. Take a look at the sketch in Figure 4.2:

Once again, we've hatched two pairs of triangles.

1. And, once again, we assert that $\triangle AMP \cong \triangle MBP$. (The sign $\cong$ is our sign for "congruent"). This is because:

a. $\overline{AM} = \overline{MB}$, since *M* is at the center.

b. $\overline{AP} = \overline{PB}$, because *P* is on the perpendicular bisector.

c. $\overline{PM}$ is the same in both triangles.

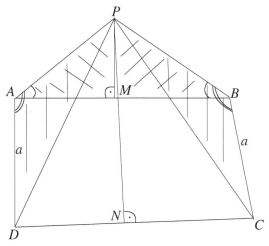

**Fig. 4.2** A first clue: The point of intersection isn't really inside the rectangle

That is, once again all three sides are identical, and thus the triangles are congruent because

$$\angle MAP = \angle MBP.$$

2. For the vertically hatched triangles, we have:

   a. $\overline{AP} = \overline{PB}$, because $P$ is on the perpendicular bisector of $\overline{AB}$.
   b. $\overline{DP} = \overline{CP}$, because $P$ is on the perpendicular bisector of $\overline{CD}$.
   c. $\overline{AD} = \overline{BC}$, by construction.

Therefore these two triangles are congruent, and thus

$$\angle DAP = \angle CBP.$$

Now, we just have to subtract the relevant angle to see that

$$90° = 100°$$

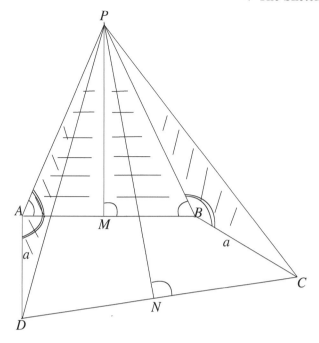

**Fig. 4.3** Sketch to prove that $90° is \neq 100°$.

## 4.5 The Complete Truth

You were heading in the right direction when you assumed that something had to be wrong with the sketch. But, in actual fact, it's worse than that. In order to demonstrate this, we'll have to increase our 100° by a lot. Nowhere in the "proof" did we use anything of the 100°. So, it could easily be 120° or more. Let's take a look at the sketch in Figure 4.3.

The point of intersection now is actually well above the figure. It's located so far above that the line connecting $P$ to $C$ actually runs **outside** the figure.

Our argument was in fact correct; we just applied it to the wrong sketch.

We've demonstrated that the two small angles at $A$ and $B$ are identical; we've also shown that the two angles indicated by double arcs at $A$ and $B$ are identical as well. But we didn't say anything about the $100°$ angle, which we should have looked at.

That was bad, right?


## 4.6 The Moral

So, what have we learned from this task? It's easy to be misled by a drawing. Sketches can be very helpful when we are attempting to discover new facts. But they can't be used to prove anything. For that, we need to think.

I'd like to summarize the result of analyzing this neat problem in a handy little rule.

The moral of the story is:
Don't trust a sketch!

# Chapter 5
# The Parallel Parking Problem

## 5.1 Introduction

Who hasn't run into the problem of trying to find a parking space in a crowded city? And then you suddenly see a free space on the side of the street, but you don't know if you'll get into it.

Such a problem is a great task for a mathematician. An everyday physical situation wants to be analyzed. So, we first have to create a mathematical model which reflects the essential features of the problem. That's normally possible only with considerable simplification, which doesn't surprise the physicist or the engineer, because they've been doing this for many centuries now.

To start with, we'd like to present a solution to this problem which created quite a stir in the German media in the fall of 2003. It was reported in many radio and television programs. The author was happy because so many people became interested in mathematics and asked him many questions. Maybe, dear reader, you'll acknowledge that mathematics is not just a means to calculate, but also offers solutions to many everyday problems which surprise even the experts sometimes. You just need to take a look at the result (8.12) for the skidding problem studied in Chapter 8.

N. Herrmann, *The Beauty of Everyday Mathematics*,
DOI 10.1007/978-3-642-22104-0_5, © Springer-Verlag Berlin Heidelberg 2012

## 5.2 The Problem

We've all been tortured with parallel parking when learning to drive. This might help here, because that's what this chapter is all about. How did it go again?

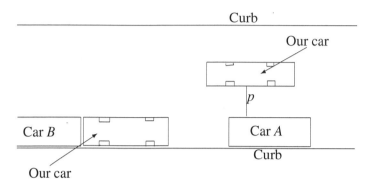

**Fig. 5.1**  There's a free space. You just need to back into it.

1. You stop directly next to the car in front, $A$, at a distance $p$ from that car (Figure 5.1).

2. You back up until the midpoint of your car, between the four tires, is parallel with the rear end of the car in front, $A$. (Later, we'll see that this can be improved. You just need to back up until the rear axle of your car coincides with the rear end of the neighboring car.)

3. Now, you have to turn the steering wheel until it doesn't turn anymore, so that you can drive into the space. In so doing, you have to drive along a circular arc with an included angle $\alpha$.

4. Then, you turn the front wheel as far as possible in the opposite direction in order to be once again parallel to the street and drive in an oppositely curved circular arc with the same included angle $\alpha$.

5. Finally, you drive forward a bit so as not to wedge in the car $B$ behind.

The following questions now arise:

---

**The Parallel Parking Problem**

1. How big does the parking space (of length $g$) have to be so that you can get into it?
2. At what distance $p$ does this maneuver begin?
3. What circular arc do you need to follow? In other words, how big is the angle $\alpha$?

---

## 5.3 Rebecca Hoyle's Formula

In mid April of 2003, the following "formula for parking" appeared around the world in many online newspapers (for typesetting reasons, it's written here on two lines):

$$p = r - w/2, g) - w + 2r + b, f) - w + 2r - fg$$
$$\max((r+w/2)^2 + f^2, (r+w/2)^2 + b^2) \pounds \min((2r)^2, (r+w/2+k)^2)$$

Internet users who happened to find this formula were confused because the formula did not make much sense.

1. Why have a closed bracket right at the beginning when there hadn't been any previous open brackets?

2. Why are there so many commas in the middle?

3. What does the British pound sign in the second line refer to?

Supposedly, this formula came from the British mathematician Rebecca Hoyle. Well, in all fairness to her, let it be said that she had not published such an absurd formula. On her website, you'll find the following formula system, which consists of four separate formulas:

$$p = r - \frac{w}{2}, \tag{5.1}$$

$$g \geq w + 2r + b, \tag{5.2}$$

$$f \leq w + 2r - fg, \tag{5.3}$$

$$\max \left( \left( r + \frac{w}{2} \right)^2 + f^2 \, , \, \left( r + \frac{w}{2} \right)^2 + b^2 \right).$$

$$\leq \min \left( 4r^2, \left( r + \frac{w}{2} + k \right)^2 \right). \tag{5.4}$$

In these formulas,

- $p$ is the lateral distance from the car in front, $A$;
- $r$ is the radius of the smallest circle which can be created by the center of the car, i.e. the center of the rectangle consisting of the four tires;
- $w$ is the width of our car;
- $g$ is the size of the parking space;
- $f$ is the distance from the car's midpoint to the front of the car;
- $b$ is the distance from the car's midpoint to the rear of the car;
- $fg$ is the distance from the car in front at the end of the parking maneuver;
- $k$ is the distance from the curb at the end of the parking maneuver.

This explains the above absurd formula. Every comma separates two formulas. Then, apparently, some word-processing program didn't understand the mathematical symbol $\geq$ and simply turned the symbol $>$ into a bracket ) while also turning the half equals sign into a subtraction. It's not clear, though, how the $\leq$ sign became the British pound sign £.

## 5.4 Criticizing Hoyle's Formula

Hoyle's formulas are, in principle, mathematically sound, and you can actually derive numerical values which describe parking.

But we also have to add a few points of criticism.

1. In the formula (5.3), only $fg$ is unknown and can be calculated. But we have to ask the significance of this distance from the car in front is when parking. This formula appears to be superfluous.

2. The same applies to the formula (5.4), which allows you to calculate the unknown value of the distance $k$ from the curbstone at the end of the parking maneuver; but you have to also ask yourself, why do you need it? This (correct) formula is also not needed for the parking maneuver.

3. The formula (5.1) seems to be useful because it provides the lateral distance that you have to keep from the car in front at the beginning.

4. The Formula (5.2) appears to be the main formula. It describes how big the space has to be in order to successfully complete the parking maneuver.

If we apply the above formulae to a typical midsized car, we see that not all of them really work. To recognize this, we need to analyze the parking maneuver really closely.

## 5.5 The Turning Circle

It's probably clear to everyone that we would drive around in circles in a large parking lot if we were to turn the steering wheel as far as possible in one direction (Figure 5.2).

We tried it and were very astonished to discover that the car didn't return exactly to the starting point after 360° but actually reached at a point about 30 cm from it. This process continued on its merry way: after every circuit, the car was somewhere else. That was, of course, due to the surface, which was not quite even. You can't really turn the steering wheel as tightly as you'd like, because the tires flex. A number of different factors come together so that you can speak only with some reservations of a true circle. Although you actually find the diameter of a turning circle listed in the documents about a car, you'll only find, for example, a turning-circle diameter of 11 m for a Golf. Not 10.952 m, but a round number of 11 m. All other measurements of a vehicle are specified exactly to the millimeter. But it's not possible to be more precise about the diameter of a turning circle. OK, we'll

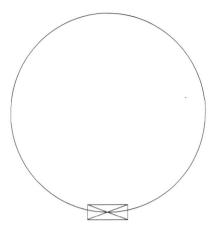

**Fig. 5.2**  Assumed turning circle of a car.

just have to accept that. For our mathematical model, though, we'll assume a perfect circle and a diameter that can be stated precisely.

After having checked with a number of different car companies, we can reveal that the turning circle of a vehicle is determined as follows:

**Definition 5.1** *If you attach a rod that drags on the ground to the right front corner of a vehicle and then drive the vehicle the smallest possible left-hand circle with the steering wheel turned as far as possible to one side, this rod will mark a circle on the ground, which is the turning circle of that vehicle. Its diameter D is entered into the documentation of the car.*

*We denote the radius of the turning circle by R; that is, half the diameter.*

If the car is well built, you can also drive in a right-hand circle and attach the rod to the left front side. Symmetry will, theoretically, lead to the same result.

## 5.6 The Center of the Turning Circle

Mathematicians immediately ask for the radius when they encounter a circle, such as the one we have just defined, and the center. Where, then, is the center of the turning circle?

Our first idea might be to find the center of the vehicle by drawing two diagonals. We then subtract the radius of the circle and we've got the center of the circle. Could be, but let's stop and think. The car is not symmetrical as far as its driving is concerned. The front wheels can be turned, while the rear wheels are rigid. Now imagine the car being driven along the turning circle. If we consider the tires to be small straight lines, then these will roll parallel to tangents to this circle. (I hope you can see this.) That's when we remember the following fact from school:

**Theorem 5.1** *The tangents to a circle are perpendicular to the radius.*

This means that we have to go perpendicular to the rear wheels, i.e., in the direction of the rear axle to find the center. The center of the turning circle is directly on the extension of the rear axle (Figure 5.3).

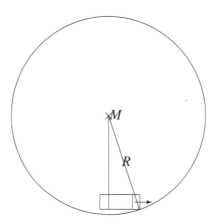

**Fig. 5.3** Correct turning circle of a car, with center.

## 5.7 The Smallest Possible Circle

Now, we've done quite a bit already with our parking model. There's now an important question. Where do we have to stop and then to start the parking maneuver? We don't want to damage the neighboring car. We saw that when we turn left, our right front fender actually defines the turning circle. But our left front wheel also turns in a circle, as do both rear wheels. For that matter, every point of the car turns in a circle. These circles do not intersect with one another; they're all concentric, with a common center $M$. Now, we ask:

Which point of our car follows the smallest circle?

If we go through our parking maneuver in such a way that the point following the smallest possible circle doesn't touch the neighboring car, then we're on the safe side. We can park without any dents. Which point, then, is so important? Take a look at the sketch in Figure 5.4, to which we have added a number of circles.

I hope you're not losing sight of what's going on here, but instead recognize with your eagle eyes that the outermost point of the left rear wheel has the smallest circle. We've added this circle to the sketch. So, we'll have to remember this point. We now have a rule for starting the parking maneuver:

We position the car next to the neighboring car in such a way that our rear axle is exactly in line with the neighbor's bumper.

Now, we can just turn the steering wheel and back up because we won't damage the neighboring car. At this point, a lot of different suggestions are made by driving instructors. For example, the right-hand door tiller, the rear seat, and even a point on the outside mirror are all mentioned in this context. Stopping and thinking about it should tell us what is the right point: the rear axle of our car, in line with the bumper of our neighbor.

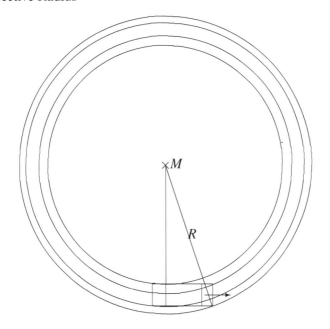

**Fig. 5.4** The smallest circle is followed by the left outermost point of the rear axle.

## 5.8 The Effective Radius

Hoyle used a radius $r$ in her formula which, though, is not the turning radius. Let's take a look at the sketch in Figure 5.5.

We see that we can ascertain fairly easily the relationship of Hoyle's $r$ to our $R$ using the Pythagorean Theorem. This gives us the equation

$$r^2 + \left(\frac{w}{2}\right)^2 = R^2,$$

from which it follows that the effective radius $r$ is given by

$$r = \sqrt{R^2 - f^2} - \frac{w}{2}, \qquad (5.5)$$

where $f$ is the distance from the rear axle to the front, of the car, and $w$ is the vehicle's width.

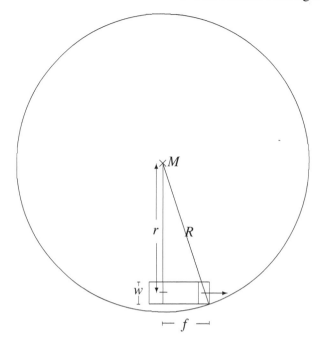

**Fig. 5.5** The relationship of $r$ to $R$

## 5.9 Our Model Car

And now to our car. Let's assume that it has the following dimensions
- we'll use round numbers for simplicity's sake:

- turning circle $D = 12$ m, and thus $R = 6$ m;
- distance between rear axle and front of car $= 3$ m;
- distance between rear axle and rear of car $= 1$ m;
- vehicle width $w = 1.5$ m.

This results in

$$r = \sqrt{R^2 - f^2} - \frac{w}{2} = \sqrt{36 - 9} - 0.75 = 4.45 \text{ m}.$$

We'll therefore calculate Hoyle's formula with $r = 4.45$ m. Then

$$p = r - \frac{w}{2} = 3.69 \text{ m},$$

$$g \geq w + 2r + b = 1.5 + 2 \cdot 4.45 + 1 = 11.40 \text{ m}.$$

This means, therefore, that we have to keep a lateral distance of 3.69 m from the neighboring car. But Dr. Hoyle, if the street is only 6 m wide, then either you're parked smack in the middle of the oncoming traffic or you're in the car that's parked on the other side of the street.

If you see a parking space 11.40 m anywhere, then you'll enter it forwards, because there is enough room for two cars. So we don't need our maneuver at all.

That couldn't be the intention. Hoyle's formulas are correct, but don't have enough precision. And that's problematic in European cities, with their narrow streets.

## 5.10 New Formulas for Parallel Parking

We need new and better formulae. Let's take a look, therefore, at the sketch in Figure 5.6.

For the derivation of our formulae, we'll use a simple trick: We look at *exiting the parking space*, not entering it. We thus go from the inside to the outside, because we can then start from a fixed position that is located directly behind the rear car. Where we end up is where our actual starting point is located, because the entire procedure is fully reversible.

So, let's take a look at our sketch. Our car in the middle of the parking space, close to the curbstone. To exit the parking space, we first back up until we almost touch the rear car. Mathematically, we'll take this distance as 0 mm. Please keep in mind, though, that this is only the theory; in the real world, the distance should be around 5 cm. Then we turn the steering wheel as far as possible to the left and start driving carefully. We drive along a circular arc with an angle of $\alpha$.

We'll leave this angle in such general terms for now because we want to create a general formula. Later, we'll have to specify it more exactly.

Once we've passed through this arc, we then stop and turn the steering wheel as far as possible in the other direction so that we once again drive along a circular arc with an angle of $\alpha$. We're now outside the parking space and parallel to the curbside, because we've driven twice along circular arcs with the same angle.

Now comes a little bit of geometry. Take a look again at the sketch in Figure 5.6. In this figure, we've included the distance $d$ from our car to the curbside and the distance $x$ from the midpoint of the rear axle to the projection of the point of contact of the two turning circles on to the perpendicular radius. If we extend the projected line to the right all the way to the perpendicular radius drawn through the right car, then the symmetry of the configuration tells us that the distance from the center of our car at the top right to this projected line is also exactly $x$. The total distance is thus $2x$. If we now also move the line toward the bottom by half the width of the car, $w/2$, we land exactly at $d$. For this reason the distance $d$ from our car to the curbside equals $2x$. We now calculate $x$ via the cosine of the angle $\alpha$:

$$\begin{aligned} d = 2x = 2 \cdot (r - y) &= 2 \cdot (r - r \cdot \cos \alpha) \\ &= 2r \cdot (1 - \cos \alpha). \end{aligned} \tag{5.6}$$

The distance $p$ from the neighboring car is

$$p = d - w = 2r \cdot (1 - \cos \alpha) - w. \tag{5.7}$$

For the length of the parking space, we get

$$g \geq 2r \cdot \sin \alpha + b. \tag{5.8}$$

If you compare the size of this space with Hoyle's formula (5.2), then you'd guess that Hoyle wants to move through an angle of $90°$, i.e. a quarter circle.

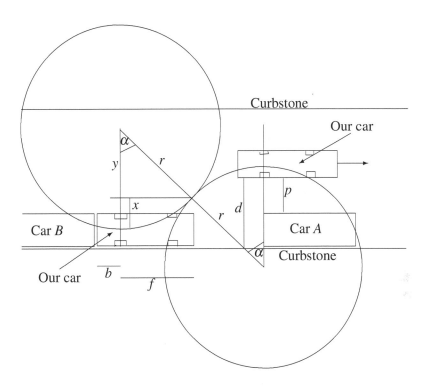

**Fig. 5.6** The starting and end positions of a parallel-parking maneuver. Our car is initially parallel to the car in front at a distance $p$, with the rear axle in line with the rear bumper of the car in front. It then moves along two circular arcs with the same angle $\alpha$ into the parking space, all the way to the end position.

## 5.11 The Formula for a 45 degree Maneuver

In some driver education classes, they tell you to drive an eighth of a circle; this corresponds to an angle $\alpha = 45°$. The formulae actually look quite simple, because $\sin 45° = \cos 45° = \frac{1}{2} \cdot \sqrt{2}$; in other words,

$$p = 2r \cdot \left(1 - \frac{1}{2}\sqrt{2}\right) - w = r(2 - \sqrt{2}) - w, \qquad (5.9)$$

$$g \geq 2r \cdot \frac{1}{2}\sqrt{2} + b = \sqrt{2} \cdot r + b. \qquad (5.10)$$

For our standard car (see Sect. 5.9), we get the following values:

$$p = 1.11 \text{ m}, \quad g \geq 7.29 \text{ m}.$$

That's still too big. No reasonable driver would stop at a lateral distance of 1.11 m from neighboring car in the street. You can imagine the honking horns that would cause.

## 5.12 The Optimal Formulas

From the formula (5.8), we see that the required size of our parking space depends a lot on the angle $\alpha$: the smaller the angle that we select, the smaller the space needed. We get the best result if we stop directly next to the car in front. When $p = 0$, i.e. the distance from the neighboring car is zero, we get the optimal angle $\alpha$ that solves our parking problem:

$$\alpha = \arccos \frac{2r - w}{2r}. \tag{5.11}$$

For our standard car, this results in

$$\alpha = \arccos \frac{2 \cdot 4.45 - 1.5}{2 \cdot 4.45} = 34°.$$

Our required space can actually be even smaller if we assume that we first reverse quite some distance when exiting the parking space and then just miss hitting the car in front of us. That's why this car can stand a bit closer.

The length of the smallest possible parking space thus is given by

$$g \geq \sqrt{2r \cdot w + f^2} + b, \tag{5.12}$$

which means

$$g \geq \sqrt{2 \cdot 4.45 \cdot 1.5 + 9} + 1 = 5.73 \text{ m}$$

for our standard car. That looks quite reasonable. In the German regulations for road construction, specifically for parking spaces, you can find a specification that a parking space on a roadside has to have at least a length of 5.75 m if it is to be used for parallel parking. That seems a bit tight for our car, but we don't really need all of this parking space. We just need room to maneuver. Our neighbors to the front and rear need this as well, so a little bit of their space can be used by us. That has obviously been taken into account in formulating the regulation.

## 5.13 Conclusions

A few concluding remarks need to be made here.

In principle, we've calculated two sizes for a parking space, namely (5.8) and (5.12). Normally, (5.8) gives a larger space than (5.12).

We now summarize our results in the following box.

---

**The new formulas for parking**

Distance from neighboring car,     $p = 0$.

Angle of circular arc,                   $\alpha = \arccos \dfrac{2r - w}{2r}$.

Required length of parking space,  $g \geq \sqrt{2rw + f^2} + b$.

---

Naturally, these formulae are products of theory, and thus are not yet applicable in the real world.

- A distance of 0 mm from the neighboring car can only be maintained by a theoretician. In the real world, more realistic values have to be used. That's why all the other values have been calculated in such a way that one doesn't hit the neighboring cars. All drivers are thus advised to include a certain leeway.

- Who wants to go hunting for a parking space with a tape measure and a protractor? It seems at least conceivable that a car designer might get the idea of feeding the necessary data into the car's computer and thus have the entire parking maneuver carried out automatically.

- Cars with a very long hood could run into trouble if the parking maneuver described here is used. With the starting position specified above, we'd crash into the car in front during the second circular arc as we move into the space. And in the final position, we'd have too much room to the rear.

  Well, we've calculated the size of the parking space in such a way that we can exit it. And if we can get out, we can also enter! The problem is only in the starting position.

  The room that we have to the rear at the end is something that have to consider at the beginning. That's why you don't start with the rear axle parallel to the rear bumper of the car in front, but with the remaining distance that you have to the rear at the end of the parking maneuver.

- Returning to a comment made above, however, you won't know the actual surplus distance to the rear until you've entered the parking space. That is, after you've given the car in front a really big dent. But that's not what we want to do. That's why a computer has to calculate everything in advance.

## 5.14 Values for a Few Cars

In Table 5.1, we've compiled some values from the drivers' manuals of a few European cars. But the interested reader should keep the above closing remarks in mind!

| Make | Model | Angle | Minimum Space (m) |
|---|---|---|---|
| VW | Golf 4 | 42° | 5.50 |
| BMW | 3 series E46 | 45° | 5.96 |
| Mercedes | C class | 43° | 5.88 |
| Mercedes | A class | 40° | 5.27 |
| Audi | A4/S4 limousine | 43° | 5.94 |
| Opel | Astra 5-door | 43° | 5.42 |
| VW | Passat | 41° | 6.08 |
| VW | Polo | 41° | 5.29 |
| Ford | Focus limousine 4-door | 42° | 5.87 |
| Mercedes | E class | 42° | 6.27 |
| Opel | Corsa limousine 3-door | 42° | 5.16 |
| Mercedes | Smart | 42° | 4.00 |
| Renault | Laguna | 46° | 6.14 |

**Table 5.1** Angles and spaces to park some European cars.

The author's website www.ifam.uni-hannover.de/~herrmann has a Java applet which calculates the angle and the length of the space required to park any car.

## 5.15  A Little Mental Exercise

Take a look at the following sequence of symbols. Do you recognize
the formation law?

$$\infty, \ \mathrm{M}, \ \Omega, \ 8, \ \text{M}\!\!\!\!/, \ \overline{6}, \ \ldots$$

Maybe you need a clue? Even first-graders will recognize the solution. Still not enough tips?

In that case, here are a few more members of the sequence which,
though, don't come across so nicely in LaTeX. But they ought to make
the answer clear.

$$\partial 6, \ \boxminus$$

Still not clear?

Well, then, take a piece of paper and cover the left halves of the
symbols. Finally got it? You might want to add a couple of symbols
for your friends.

# Chapter 6
# The Parking Garage Problem

## 6.1 Introduction

So, we've learned a lot about parking in the last chapter if there's a free parking space parallel to the curbside. But what about a parking garage or a street where the parking spaces are perpendicular to the road?

This is a simple task, which we can solve with just a little help. We'll use this help here, because it will also give those readers who are not so well versed a chance to apply their mathematical insights.

## 6.2 The Problem

First, we have to be certain of the task. What exactly is the problem? Anyone who has parked in a garage knows that you have to keep quite some distance from the parking space to the front, to the rear, and to the side. More experienced drivers have probably already tried parking backwards and, therefore, know that you can keep less distance from the side when conducting such a maneuver. This, therefore, calls for a more precise formula. But first, the problem:

N. Herrmann, *The Beauty of Everyday Mathematics*,
DOI 10.1007/978-3-642-22104-0_6, © Springer-Verlag Berlin Heidelberg 2012

---

**The Parking Garage Problem**

What's the right position to enter a parking space that's perpendicular to the road in one move, either forward or backward?

---

## 6.3 Forward Parking

Now, we need a sketch to depict the situation (Figure 6.1).

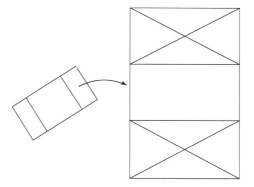

**Fig. 6.1** How lucky can you get, a free parking space! But how to enter it?

The sketch looks quite harmless. But what if you're sitting there, trying to find the right gear, while in the meantime lots of cars are backing up behind you, with a lot of angry drivers. You start sweating. But stop, let's think about this.

Our trick for finding the right solution is the same as in the previous chapter:

**We mentally put ourselves into the parking space and then leave it. That's how we reach the starting point we're looking for.**

We'll simplify the situation by assuming that we put ourselves directly in front of the empty parking space by backing out of the parking space in a straight line. This situation is shown in Figure 6.2. In a real-life situation, we'd start turning the wheel a bit earlier; this allows us to save a couple of inches. But we'll leave this maneuver to the more experienced readers. After all, we just want to explain the principle here.

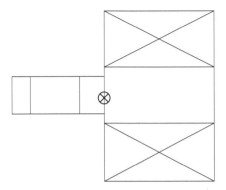

**Fig. 6.2** We place ourselves directly in front of the free parking space by backing out in a straight line so that the middle of our front bumper is exactly at ⊗.

Here, we turn the steering wheel as far as possible – the direction depends on which way we want to go – and continue to back up along a complete quarter arc. This corresponds to an angle $\alpha = 90°$. And that's how we reach the position from which we have to start in order to enter the parking space. Figure 6.3 shows us exactly this position.

We now specify the coordinates for the points marked ⊗ so as to be able to describe our position exactly. We take the point in the middle of the entrance to the parking space as our zero point; that is, the point with coordinates $(0,0)$. Consequently, the point that marks the exact center of the front end of our car has coordinates $(-r-f, r-f)$. Here we're using the same notation as in Chapter 5:

- $r$ is the effective turning radius according to the formula (5.5),

$$r = \sqrt{R^2 - f^2} - \frac{w}{2};$$

- $f$ is the distance from the rear axle to the front of the car.

Let's put this in a more understandable way. From the midpoint of our desired parking space, we need to keep a distance of $r + f$ from the side while also having some room for our hood, $r - f$, at the front. To make things even clearer, we'll take the same cars as in the previous chapter and list the distances they have to maintain (Table 6.1).

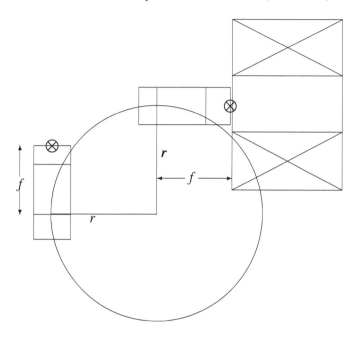

**Fig. 6.3** On the left, we have our final position after we have left the parking space; that is, the starting position for parking.

Whenever a negative distance from the (center of the) parking space is listed in the table, you need to put the hood beyond the midpoint of the parking space by the specified distance and then start turning the wheel. Some cars just have an extremely long hood.

## 6.4 Backward Parking

Through experience, we've learned that we can start from a different position when parking backwards. A sketch (Figure 6.4) will once again explain why exactly this is so.

| Make | Model | distance (m) from the side | distance (m) from parking space |
|------|-------|----------------------------|---------------------------------|
| VW | 4th-generation Golf | 6.76 | 0.10 |
| BMW | 3-series E46 | 6.55 | −0.63 |
| Mercedes | C class | 6.68 | −0.24 |
| Mercedes | A class | 6.63 | 0.63 |
| Audi | A4/S4 limousine | 6.93 | −0.21 |
| Opel | Astra 5-door model | 6.55 | −0.29 |
| VW | Passat | 7.15 | −0.17 |
| VW | Polo | 6.61 | 0.09 |
| Ford | Focus 4-door limousine | 6.82 | −0.06 |
| Mercedes | E class | 7,20 | −0.18 |
| Opel | Corsa 3-door limousine | 6.49 | −0.071 |
| Mercedes | Smart | 5.07 | 0.74 |
| Renault | Laguna | 7.14 | −0.30 |

**Table 6.1** Distances from the side and from the parking space required for forward parking of several European cars.

As before the notation is as follows:

- $r$ is the effective turning radius according to the formula (5.5),

$$r = \sqrt{R^2 - f^2} - \frac{w}{2};$$

- $b$ is the distance from the rear axle to the back of the car.

And there you have it. The distance from the side is $r + b$, instead of $r + f$ as before.

From the midpoint of our desired parking space, we need to keep a distance of $r + b$ from the side, while our trailer hitch needs to keep a distance of $r - b$ from the rear of the space.

The distance from the side is thus smaller than before since $b$ is much smaller than $f$. However, the distance from the parking space has to be a bit bigger than before, namely $r$ instead of $r - f$.

This is because the center of the turning circle is directly in line with the rear axle. So when we park backwards, we can take a position closer to the parking space, much to the delight of the drivers behind

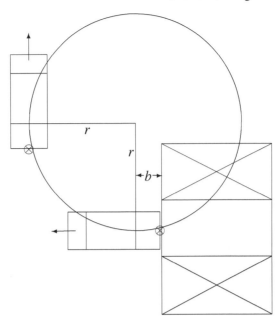

**Fig. 6.4** We place ourselves directly in front of the free parking space by driving out straight out of the space so that our trailer hitch is at ⊗.

us, because they can squeeze past us to the side while we're looking for reverse. The slightly bigger distance from the parking space merely requires us to use the indicators to tell the drivers behind us that we're about to enter "our" parking space, so that they don't try to steal it.

The distances required for several European cars are listed in Table 6.2.

| Make | Model | distances (m) to the side | distances (m) to parking space |
|------|-------|---------------------------|--------------------------------|
| VW | 4th generation Golf | 6.76 | 0.10 |
| BMW | 3-series E46 | 6.55 | −0.63 |
| Mercedes | C class | 6.68 | −0.24 |
| Mercedes | A class | 6.63 | 0.63 |
| Audi | A4/S4 limousine | 6.93 | −0.21 |
| Opel | Astra 5-door model | 6.55 | −0.29 |
| VW | Passat | 7.15 | −0.17 |
| VW | Polo | 6.61 | 0.09 |
| Ford | Focus 4-door limousine | 6.82 | −0.06 |
| Mercedes | E class | 7, 20 | −0.18 |
| Opel | Corsa 3-door limousine | 6.49 | −0.071 |
| Mercedes | Smart | 5.07 | 0.74 |
| Renault | Laguna | 7.14 | −0.30 |

**Table 6.2** Distances from the side and from the parking space required for backward parking of several European cars.

# Chapter 7

# The 85th Birthday Problem

My mother-in-law recently turned 85. Of course, that was a big event where math participated as well. In particular, I had the opportunity of talking a little bit about math in front of a large audience of very interested guests. One doesn't always get such an opportunity. But what should I tell them? Well, I used a few good examples to explain the beauty of math. This is my speech.

## 7.1 Dear Mother-in-Law

Today, we're celebrating your 85th birthday. But stop! The mathematician hesitates. Was the day you were born your first birthday, or was it actually your zeroth birthday? Well, we've always started to count with 1. That's the reason why the first year of our calendar is the year 1 after the birth of Christ, i.e. 1 A.D. The year before that year was the year 1 B.C. Historians don't know the year zero.

And that's exactly why the new millennium should actually have been celebrated on January 1, 2001. This is substantiated by a very simple argument:

> If a crate of beer contains 20 bottles, then the twentieth is, unfortunately, the last one. A new case begins with the 21st bottle. Isn't that fantastically simple?

But be careful, it's different with astronomers. They count the year 0 directly before the year +1, and before that the year −1. So, if you

N. Herrmann, *The Beauty of Everyday Mathematics*,
DOI 10.1007/978-3-642-22104-0_7, © Springer-Verlag Berlin Heidelberg 2012

want to ask an astronomer about the night sky in the year 333 B.C. (the year of the Battle of Issus), you have to ask about the night sky in the year −332. Isn't that strange?

But let's get back to counting your birthdays, dear mother-in-law.

When you were one year old, that was already your second birthday. So, according to this method of counting, it's your 86th birthday today. That's why we should actually be saying:

*Today, we're celebrating the highly praised 85th return of your birthday.*

And you're completing 85 years today. We're actually celebrating 85 full and wonderful years with you.

How can we commemorate this event mathematically?

## 7.2 What Do Mathematicians Do?

Well, first we have to pursue the question of what mathematicians actually do. All non-mathematicians seem to know this. At least, I constantly hear this when I identify myself as a mathematician:

*Oh, I've never been able to do math.*

Do people really think mathematicians would multiply 17 by 49 all day long? That's not true! I usually answer with a provocative assertion:

*Then we have something in common. Because I also can't do math very well.*

That causes the other person to really start thinking. How can that be, a mathematician who can't do math? He's just showing off!

If you look at real math books, you'll see that they actually contain numbers. We can find page numbers, and we also like to number our theorems and definitions so that it's easier to refer to them. But other than that, numbers are used only in examples. In order to verify our theories, it's necessary to calculate from time to time. But these calculations are just examples.

That's why numbers belong to math even though they're not vital.

## 7.3 The Numbers of Your Life

Two numbers, dear mother-in-law, belong to your life because they are an essential part of it.

- First of all, there is the time of your birth. That's the beginning and, thus, zero.
- And second, today's very important number 85.

Both of these numbers are very interesting from a mathematical point of view. However, that is not a very profound statement, because from a mathematical point of view, all numbers are interesting. You don't believe me?

Well, let's assume that there are natural numbers that are not interesting. This set of not-interesting natural numbers contains a smallest number, of course; possibly 1, the smallest natural number. This number would be

the smallest uninteresting number.

Oh, but I beseech you, that would truly be a very interesting number. It can't really belong to the uninteresting numbers. We have to take it out of that category. That leaves the remainder of the uninteresting numbers. Among them is, once again, a smallest one. We can draw the same conclusion again. Because this number would once again be very interesting, as the smallest uninteresting number, etc., etc.

At the end of the process, we'll have taken every number out, so that no number that uninteresting remains.

## 7.4 The Number Zero

Let's start with zero.

### *The Definition of Zero*

As mathematicians, we need to be sure of what we are discussing. So, then, what is zero?

**Definition 7.1** *By the zero, represented by the symbol* 0, *we mean the number which doesn't change anything when it is added; in other words,*

$$0 + a = 0 \text{ for all numbers } a. \tag{7.1}$$

### *The Uniqueness of Zero*

Zero doesn't hurt us. Wait a minute, did we talk about *the* zero? That was a bit hasty, because there might actually be lots of zeros. Think about it!

Let's assume that in addition to the 0 introduced above, there is another number with the property (7.1). Let's call it $\widetilde{0}$ to differentiate it from 0. We thus have the following property for $\widetilde{0}$:

$$\widetilde{0} + a = \widetilde{0} \text{ for all numbers } a. \tag{7.2}$$

We now think hard. Let's add these two "zeros":

$$0 + \widetilde{0}.$$

As we know, addition of the 0 won't change the result, so we get

$$0 + \widetilde{0} = \widetilde{0}. \tag{7.3}$$

But this is exactly the same property that $\tilde{0}$ has. By the same argument, we can also show that

$$0+\tilde{0}=0. \qquad (7.4)$$

If we now put together (7.3) and (7.4), we get

$$0=0+\tilde{0}=\tilde{0}, \qquad (7.5)$$

and therefore

$$0=\tilde{0}. \qquad (7.6)$$

Thus, there is only one number 0. If we find a calculation somewhere where nothing is changed by the addition of a term, then this term is definitely zero. We'll apply this knowledge in the next subsection.

### Multiplying by Zero Gives Us Nothing

We have learned that we can add zero without hesitation anywhere. It simply doesn't change anything. What happens, though, when we multiply by zero?

From our childhood, we know the following rule:

$$0\cdot a=0. \qquad (7.7)$$

Is this a definition, or can we derive it?

We can actually prove it. This is done with a neat little trick. We use the equation

$$0+0=0, \qquad (7.8)$$

which is derived from the definition of 0. Next, we do the following calculation:

$$0 \cdot a = (0+0) \cdot a = 0 \cdot a + \underbrace{0 \cdot a}. \tag{7.9}$$

We have to really savor this equation. What is on the right-hand side of the equation? We have added something to the number $0 \cdot a$ that is unknown to us at this point, namely $0 \cdot a$, and nothing has changed because the left-hand side is $0 \cdot a$. According to our considerations above, this is possible only if we have added zero. So, the part with a brace underneath is zero; that is,

$$0 \cdot a = 0, \tag{7.10}$$

QED.

### Division by Zero

Now we know how to do almost all calculations involving zero. Well, we can see right away that we can subtract it without changing anything. So, that leaves only division. What do we get when we divide by zero?

There are a number of wise sayings, such as:

Dividing by zero is the biggest epic failure known to mankind!

Well, what's so bad about dividing by zero?

Let's think about it:

- When I divide 10 by 1, I get 10.
- When I divide 10 by $0.5 = 1/2$, I get 20.
- When I divide 10 by $0.25 = 1/4$, I get 40.
- When I divide 10 by $0.1 = 1/10$, I get 100.
- When I divide 10 by $0.01 = 1/100$, I get 1000.
- When I divide 10 by $0.000001 = 1/1000000$, I get 10,000,000.

So, when I divide by increasingly smaller numbers, I always get something bigger than the previous result. One could thus assume:

When I divide by zero, I get infinity, represented by the symbol ∞.

If we want to permit division by zero, we also have to calculate with infinity, that is, ∞.

### Hilbert's Hotel

Here's a nice little story.

David Hilbert (1862–1943), one of the best mathematicians of the twentieth century, owned a hotel, the famous Hilbert's Hotel. It was famous because it had an infinite number of rooms. One evening, what luck for Mr. Hilbert, all rooms were occupied. But it so happened that Carl Friedrich Gauß (1777–1855), the Princeps Mathematicorum, appeared and wanted a room. What to do?

Hilbert couldn't turn the famous Mr. Gauß away. Fortunately, he had the following idea:

- He asked the guest in Room No. 1 to move into Room No. 2.

- He asked the guest in Room No. 2 to move into Room No. 3.

- He asked the guest in Room No. 3 to move into Room No. 4.

- etc.

Consequently, all of the guests who had so far been staying in the hotel had a room once again; but, surprisingly, Room No. 1 became vacant and Mr. Gauß was able to move in.

The moral of the story is

$$\infty + 1 = \infty. \tag{7.11}$$

But this equation should immediately give us cause to stop and think. Something has been added to the number ∞; in this case, 1. And everything stays as it is, because the result is once again ∞. But we discovered above that only the number zero has this no-change property. So, we can conclude that

$$1 = 0.$$

So, if we recognize $\infty$ as a genuine number because we want to divide by zero, it follows that $0 = 1$. That is a truly absurd equation.

This clearly means that we can't count on $\infty$ to behave as a number. And this means, in line with our above considerations:

*We're not permitted to divide by zero.*

**Remark 7.1** *If we want to solve an equation, we're allowed to change the equation, and maybe simplify it, by doing the same thing on both sides. For example, we can add 5 or divide by 17 on both sides. But there are some special considerations for zero.*

*We made it clear above that we're not permitted to divide by zero.*

*However, we're also not allowed to multiply by zero on both sides, because we would then reduce both sides to zero. While the result would be a correct equation, it wouldn't really say anything. We couldn't get draw more conclusions from such an equation. That's why zero is something special.*

## 7.5 The Number 85

Well, dear mother-in-law, we now come to the second number that is so important for you today; namely, 85. I've prepared a little card game.

### Card Game

At home, I wrote something on four cards. I applied two rules so it didn't become too boring.

1. I wrote letters on one side of the card, and numbers on the other side.

2. When I wrote a vowel A, E, I, O, or U on one side, I wrote an even number on the other side.

Now, that's a rule that should warm your heart. Let me show you my four cards, which I've shuffled a bit. (Figure 7.1).

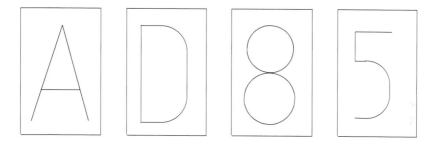

**Fig. 7.1** On two cards, we see the letters A and D; on the other two, we see 8 and 5.

On two cards, we see the letters A and D, the initials of your name, dear mother-in-law. On the other two, we see 8 and 5; in other words, your special number today.

Well, you've advanced quite a bit mathematically and have some doubts. You're worried that I might be fooling you. Maybe I've made a mistake in following the rules. In any case, we won't leave things unchecked; we'll check everything. But now the decisive question:

Which of the cards do we have to turn round to see whether I've stuck to the rules?

All of us will probably agree immediately that we have to turn round the card with the A on it. On the back of it, there must be an even number; otherwise, it is wrong. Okay, let's assume that it has a 4 on the back.

Well, are we done? Are we sure that everything is correct now, or do we have to turn another card round? The card with the D on it is not interesting, because we didn't say anything about consonants. Anything can be on the back as long as it is a number.

That leaves the cards with the numbers 8 and 5. Do we have to check one of them?

Let me give you an example. Take the following statement:

The bells ring at 12 noon.

Let's assume that we live somewhere where we can hear bells ringing, and that the bells always ring at 12 noon.

**Remark 7.2** *By the way, do you know why bells ring at noontime in Germany? It's a fairly martial custom. On June 29, 1456, Pope Calixtus III (1378–1458) issued a bull in which he asked all believers to pray at 12 noon a Hungarian victory over the Turks, while at the same time the church bells would be rung. He got the news on August 6 that the Hungarians had already won on June 22. Back then, this had to be done without any text messaging or Internet, merely with horses, which, quite naturally, took some time – Something we can hardly imagine today. But this crossing of the news with the bull was considered to be a sign. That's why the bells keep on ringing even today.*

Let's turn the above statement into a real, hard-bitten assertion:

When it is 12 noon, the bells ring.

Now, is it correct to say the following?

When the bells ring, it is 12 noon.

That's complete nonsense. The bells also ring in the evening at 6 p.m., or maybe on Sundays at 9:45 a.m., or when you request it for the baptism of your child.

Nonetheless, we can still somehow turn the statement around. Let's think about it, and then we'll get the right idea.

Because the following statement is also true:

When the bells do not ring, it is not 12 noon.

We'll accept this because it's also right. Let's transfer this to our letter–number game.

Our rule was:

Front, vowel $\Longrightarrow$ back, even number.

That's equivalent to the following statement, which is similar to the above statement about the bells:

Back, odd number $\Longrightarrow$ front, consonant.

Let's savor that for a moment, please.

We thus don't have to turn the 8 around, but we do have to turn the 5. If it has a vowel on the back, then I didn't stick to my rules. Okay, show the card. It has a U is on the back; in other words, I cheated.

**Remark 7.3** *If you want to play this game with another person or at some other occasion, you have to observe the following:*

1. *You have to conjure two letters from somewhere, one of which is a vowel and the other a consonant.*

2. *You then add two numbers, one of which is even and the other odd. That shouldn't be a problem.*

## 7.6 85 Is Everywhere

And, finally, I'd like the audience to help me a little. Please give me three numbers between 0 and 9. Why don't you choose your favorite number?

I hear 6; good! then 8; thanks! and 3; excellent! So, you have chosen

## 683

With this number, which you have given to me arbitrarily – it's important that it really is an arbitrary number, and not one that I chose in advance – I will now do manipulations.

We reverse the number, in other words read it from back to the front, and write

$$386$$

We subtract this number from the one above. Note that we want to work with positive numbers. If you reverse the number 129 you get the number 921. You then have to subtract the first number, 129, from the reversed number, otherwise you end up in the negative range.

So, we calculate as follows:

$$
\begin{array}{r}
683 \\
-386 \\
\hline
297
\end{array}
$$

We like this game! Next, we reverse the new number, reading it once again from back to front, and thus get

$$792$$

We now add the last two numbers together. But please don't annoy me with the question "why?" Let's just do this, and then we'll take a look:

$$
\begin{array}{r}
297 \\
+792 \\
\hline
1089
\end{array}
$$

We subtract 1 from this number, multiply the result by 5, and get

$$5440$$

Because this number is too big, we halve it, and halve again and again until we get a fraction:

$$5440 \rightarrow 2720 \rightarrow 1360 \rightarrow 680$$

$$\rightarrow 340 \rightarrow 170$$

$$\rightarrow 85$$

Wasn't that ingenious? The number 85 appears every time. Do you want to know how I did it? You wise guys, you fools! You can find out in my book, Können Hunde rechnen? (see [5])

I now like to wish my mother-in-law a wonderful birthday party and many more to come, with such interesting numbers.

## 7.7 State Capital Problem

To improve your mind and increase the number of gray cells in your brain, I would like now to give you an amusing problem.

Think about the 50 states of the United States and their capital cities. For example, the capital of Florida is Tallahassee. Now compare the two names Florida" and "Tallahassee," and you will see that they have two letters in common, "l" and "a." As another example, think about California and Sacramento. The two names have "c," "a," and "r" in common.

Now for my very important question: Is there a state in the USA for which the name of the state and the name of its capital city do not have any letter in common?

Mathematically, we are taking the set of all letters in the name of one state and the set of all letters in the name of its capital city and asking about the intersection of the two sets. We are interested in finding a state for which this intersection is empty.

A hint that may encourage you to hard thinking is this: There is exactly one state with this fantastic, unique property.

But now youre in trouble. Perhaps youre thinking of Alaska? But hold on; the capital is Juneau. There is an "a" in Juneau – ask your teacher or look at your geography textbook. Alaska is the wrong answer.

If you cant solve this little problem or you just want the solution now, you can take a look at page 98 and find the answer there, but please at least think about the problem first.

# Chapter 8
# The Slippery-Ice or Bread-Slicing Problem

## 8.1 Introduction

What on earth do slippery ice and the slicing of bread have in common? That's the question. We can even top that question and talk about a drill bit seizing up. All three problems can be solved with the same mathematical methods. That's amazing! That's exactly why mathematics is so thrilling. We'll first describe the tasks and then set about finding a solution.

## 8.2 The Problem

> ### The Slippery-Ice Problem
>
> Why do locked or spinning wheels make a car slide so easily to the side?

A question that is seldom asked in such detail. But anyone who has skidded on the road will want to know why the car doesn't just continue to plow on straight ahead; that is to say, just slide straight ahead. In that case there would be no problem. But no, the car wants to slide off to the side.

N. Herrmann, *The Beauty of Everyday Mathematics*,
DOI 10.1007/978-3-642-22104-0_8, © Springer-Verlag Berlin Heidelberg 2012

---

**The Bread-Slicing Problem**

Why do you cut back and forth with a knife
when all you want to do is cut downwards?

---

No one will probably have asked this question either. But if someone is actually confronted with this question, they'll be surprised,
laugh, start to think, and say "Why, yes! Why is that?" Often, people refer to the serrated edge of a knife in their attempts to answer.
But that's not what we mean. Everyone will have noticed that when
you slice bread, the knife cuts more easily – yes, it seems almost to
glide to the bottom by itself – if we move it back and forth. Our goal
is to explain why that is.

## 8.3 Physical Background

We'll first examine the skidding car and then come back to bread slicing. Actually, the same physical principles apply to both phenomena.

Take a body with a mass $m$. This is our car. It has a weight of
$W = m \cdot g$, where $g$ is the gravitational acceleration. We simplify the
problem by having the car driven on a perfectly straight road; naturally, this road is the $x$ axis, which is embedded in a perfect plane, the
$x$–$y$ plane. Otherwise, it'll get too complicated. We also imagine that
the car moves with a constant speed $v = dx/dt > 0$ where $t$ is the time.

Physics teaches us that frictional force $\mathbf{R}$ is parallel to the direction
of motion, that is, in the direction of $\mathbf{v} = \left( \frac{dx}{dt}, \frac{dy}{dt} \right)$, and is proportional
to the weight, in other words, the force perpendicular to the $x$ axis.
This introduces a dimensionless friction coefficient $\mu$:

$$\mathbf{R} = \frac{W \cdot \mu}{\sqrt{\left( \frac{dx}{dt} \right)^2 + \left( \frac{dy}{dt} \right)^2}} \cdot \begin{pmatrix} \frac{dx}{dt} \\ \frac{dy}{dt} \end{pmatrix}.$$

We consider two different types of forces which influence the vehicle. First, there is the force generated by the engine torque, which is applied in the direction of motion, that is, the $x$ axis. We call this force $J$. And then we have the dangerous lateral force, caused by crosswinds, slopes, lopsided roads, etc. This is denoted by $K$.

A fundamental physical law, formulated by Isaac Newton, states that

$$\text{force} = \text{mass} \times \text{acceleration}$$
$$\mathbf{F} = m \cdot \mathbf{a}.$$

The acceleration $\mathbf{a}$ is the second derivative with respect to time; that is, $\left( \mathbf{a} = \frac{d^2x}{dt^2}, \frac{d^2y}{dt^2} \right)$. We now summarize everything and keep in mind that the friction is in the opposite direction to the motion. Therefore it's given a negative sign:

$$\begin{pmatrix} J \\ K \end{pmatrix} - \frac{W \cdot \mu}{\sqrt{\left( \dfrac{dx}{dt} \right)^2 + \left( \dfrac{dy}{dt} \right)^2}} \cdot \begin{pmatrix} \dfrac{dx}{dt} \\ \dfrac{dy}{dt} \end{pmatrix} = \mu \cdot \begin{pmatrix} \dfrac{d^2x}{dt^2} \\ \dfrac{d^2y}{dt^2} \end{pmatrix}. \qquad (8.1)$$

## 8.4 The Mathematical Model

Before we take a closer look at this equation, we'll think about the initial state. We start by examining the situation at the time $t = 0$. Let this be the time at which the vehicle starts to slide sideways. At this point, it's still in the lane; in other words, it has not yet strayed from the $x$ axis. We express this mathematically as follows:

$$y(0) = 0.$$

At this point in time, the skidding to the side begins. But no change in the sideways motion has yet occurred. Mathematically, this means

$$\frac{dy}{dt}(0) = 0.$$

We call these two conditions *the initial conditions*. This is precisely the topic of this chapter, the study of what are called initial differential equations. Such equations are often found in applied mathematics. But because they are often constructed in a very complex manner, we want to focus on approximate solutions of such problems.

We now add another simplification. At least when the skidding starts, that is, for small $t$, the sideways skidding motion is much smaller than the forward motion:

$$\frac{dy}{dt} < \frac{dx}{dt} \Rightarrow \left(\frac{dy}{dt}\right)^2 \ll \left(\frac{dx}{dt}\right)^2.$$

This allows us to significantly simplify the square root in the denominator of (8.1). This square root now equals $dx/dt = v$.

Since we want to examine the sideways skidding, we're interested only in the second component of (8.1):

$$K - W \cdot \mu \cdot \frac{1}{v} \cdot \frac{dy}{dt} = m \cdot \frac{d^2y}{dt^2}.$$

A simple conversion produces the following equation:

$$\frac{K \cdot v}{W \cdot \mu} = \frac{dy}{dt} + \frac{m \cdot v}{W \cdot \mu} \frac{d^2y}{dt^2}, \qquad \text{where } y(0) = 0, \ \frac{dy}{dt}(0) = 0. \quad (8.2)$$

We have thus got an equation in which the function $y(t)$ that we seek appears, but it's somewhat hidden. Only its first and second derivatives are present. An equation containing derivatives of an unknown function (and possibly the function itself) is called a *differential equation*. It's actually quite amazing where one encounters differential equations. They allow one to describe many, if not most, natural phenomena.

To assist in the interpretation, we'd like to make some comments.

**Remark 8.1** *1. The first component of (8.1),*

$$J - \frac{W \cdot \mu}{v} \cdot \frac{dx}{dt} = m \cdot \frac{d^2 x}{dt^2},$$

*describes the normal motion of the car on the road; that is, in the x direction; μ is the coefficient of static friction. Manufacturers have been trying to keep μ as small as possible, which is achived by greasing the axles, finding a better rubber mixture for the tires, etc. μ → 0 means that the second term on the left disappears (there is no friction). We're looking for (and finding) a function that has a constant second derivative because we have said that the acceleration is constant. Without friction, the car would just keep going faster and faster.*

2. *The second component describes the lateral movement; now μ is the coefficient of sliding friction. Many are interested in making this as big as possible. The following equation is obtained when μ → ∞:*

$$\frac{dy}{dt} = 0 \Longrightarrow y = const.$$

*Because of the starting conditions, only the function y ≡ 0 remains as a possibility. This means that the car moves in a very straight line. There is no sideways sliding.*

*It can be concluded that the friction coefficient should be as small as possible in one particular direction, namely the direction of motion, while it should be as large as possible in the direction perpendicular to the direction of motion. Manufacturers are working on a solution.*

## 8.5 The Solution

We're now looking for a solution to (8.2).

We thus look at the following initial-value problem:

$$\frac{K \cdot v}{W \cdot \mu} = \frac{dy}{dt} + \frac{m \cdot v}{W \cdot \mu} \frac{d^2 y}{dt^2}, \qquad \text{where } y(0) = 0, \ \frac{dy}{dt}(0) = 0, (8.3)$$

$$y(0) = 0, \qquad \frac{dy}{dt}(0) = 0. \tag{8.4}$$

We use the Laplace transformation to find the solution. Since we can't assume that too many of our readers are familiar with the procedure, we'll present the following steps only briefly. You can skip this subsection if you wish and move straight on to the solution.

### Laplace Transformation

The Laplace Transform of (8.3) can be expressed as

$$\mathcal{L}\left[ \frac{mv}{W\mu} \cdot \frac{d^2 y}{dt^2} + \frac{dy}{dt} \right](s) = \mathcal{L}\left[ \frac{Kv}{W\mu} \right](s).$$

Taking advantage of the linearity of the Laplace operator and the differentiation theorem for Laplace transform, we obtain

$$\frac{mv}{W\mu} \cdot s^2 \cdot \mathcal{L}[y](s) - s \cdot \underbrace{y(0)}_{=0} - \underbrace{y'(0)}_{=0} + s \cdot \mathcal{L}[y](s) - \underbrace{y(0)}_{=0} = \frac{Kv}{W\mu} \cdot \frac{1}{s}, \ s > 0.$$

Here, we have indicated the initial conditions using the braces under the relevant terms. As a consequence of these conditions, some of the terms are not needed. On the right, we have applied some previous knowledge, namely $\mathcal{L}[1](s) = 1/s$.

In conclusion, this results in

$$\left( \frac{mv}{W\mu} \cdot s^2 + s \right) \mathcal{L}[y](s) = \frac{Kv}{W\mu} \cdot \frac{1}{s}.$$

We solve it this way:

$$\mathcal{L}[y](s) = \frac{Kv}{(mvs + W\mu) \cdot s^2}$$

$$= Kv \cdot \frac{1}{(mvs + W\mu) \cdot s^2}. \tag{8.5}$$

Now comes the difficult part, the inverse transformation. To do this, we split the term on the right into simpler terms. We do this with the help of partial fractions. We therefore take the following trial solution:

$$\frac{1}{(mvs + W\mu) \cdot s^2} = \frac{As + B}{s^2} + \frac{C}{mvs + W\mu}. \tag{8.6}$$

By multiplying this by a common denominator, we obtain

$$1 = (As + B)(mvs + W\mu) + Cs^2.$$

Three unknowns, $A$, $B$, and $C$, await their destiny. We have to find three equations for them somehow. With the help of the common-denominator trick, we've removed the singularity at $s = 0$. We can thus say that this equation is valid for all $s \in \mathbb{R}$. By selecting $s = 0$, $s = 1$, and $s = -1$, we get our three equations. First,

$$s = 0 \quad \Longrightarrow \quad 1 = B \cdot W\mu \Longrightarrow B = \frac{1}{W\mu}.$$

Second,

$$s = 1 \Rightarrow 1 = \left(A + \frac{1}{W\mu}\right)(mv + W\mu) + C = A(mv + W\mu) + \frac{mv}{W\mu} + 1 + C.$$

This results in

$$A(mv + W\mu) + C = -\frac{mv}{W\mu}. \tag{8.7}$$

Finally,

$$s = -1 \;\Rightarrow\; 1 = \left(-A + \frac{1}{W\mu}\right)(-mv + W\mu) + C$$
$$= -A(-mv + W\mu) - \frac{mv}{W\mu} + 1 + C.$$

This leads to

$$-A(-mv + W\mu) + C = \frac{mv}{W\mu}. \tag{8.8}$$

We subtract (8.8) from (8.7) and get

$$Amv + AW\mu + C - Amv + AW\mu - C = -\frac{mv}{W\mu} - \frac{mv}{W\mu},$$

and

$$A = -\frac{mv}{W^2\mu^2}. \tag{8.9}$$

We calculate $C$ by inserting this value of $A$ in (8.7):

$$C = -\frac{mv}{W\mu} + \frac{mv}{W^2\mu^2}(mv + W\mu)$$
$$= \frac{m^2v^2}{W^2\mu^2} \tag{8.10}$$

We now apply (8.6), together with (8.5):

$$\mathscr{L}[y](s) = Kv \left[ \frac{-\frac{mv}{W^2\mu^2}s + \frac{1}{W\mu}}{s^2} + \frac{\frac{m^2v^2}{W^2\mu^2}}{mvs + W\mu} \right]$$

$$= \frac{Kv}{W^2\mu^2} \left[ \frac{-mvs + W\mu}{s^2} + \frac{m^2v^2}{mvs + W\mu} \right]$$

$$= \frac{Kmv^2}{W^2\mu^2} \left[ \frac{-s + \frac{W\mu}{mv}}{s^2} + \frac{mv}{mvs + W\mu} \right]$$

$$= \frac{Kmv^2}{W^2\mu^2} \left[ \frac{-s + \frac{W\mu}{mv}}{s^2} + \frac{1}{s + \frac{W\mu}{mv}} \right]$$

$$= -\frac{Kmv^2}{W^2\mu^2} \cdot \frac{1}{s} + \frac{Kv}{W\mu} \cdot \frac{1}{s^2} + \frac{Kmv^2}{W^2\mu^2} \cdot \frac{1}{s + \frac{W\mu}{mv}}.$$

## 8.6 The Result

Now we have everything that we need, and we can perform an inverse transformation to get the solution:

$$y(t) = -\frac{K \cdot m \cdot v^2}{W^2 \cdot \mu^2} + \frac{K \cdot v}{W \cdot \mu} \cdot t + \frac{K \cdot m \cdot v^2}{W^2 \cdot \mu^2} \cdot e^{-\frac{W \cdot \mu}{m \cdot v}t}. \qquad (8.11)$$

To interpret this solution, we turn the exponential function into a power series and consider only the first few terms. Because

$$e^{-\lambda t} = 1 - \lambda t + \frac{\lambda^2 t^2}{2!} - \frac{\lambda^3 t^3}{3!} + \frac{\lambda^4 t^4}{4!} \pm \cdots$$

is used in the solution, we get

$$y(t) = \frac{Kmv^2}{W^2\mu^2} \left( -1 + 1 - \frac{W\mu}{mv}t + \frac{1}{2}\frac{W^2\mu^2}{m^2v^2}t^2 - \frac{1}{6}\frac{W^3\mu^3}{m^3v^3}t^3 \pm \cdots \right) + \frac{Kv}{W\mu}t$$

$$= \frac{1}{2}\frac{K}{m}t^2 - \frac{1}{6}\frac{WK\mu}{m^2v}t^3 \pm \cdots.$$

## 8.7 Interpretation of the Result

Physically, the solution above can be interpreted as follows. For a very tiny time $t > 0$, we can say that $t^3 \ll t^2$. When the skidding starts, in other words for small $t$, the second term can be ignored and the solution is

$$y(t) = \frac{1}{2}\frac{K}{m}t^2. \qquad (8.12)$$

And now comes the revelation:

> **There is no $\mu$ in this function.**

This means that the start of the skidding process is no longer influenced by sliding friction. Frictionless drifting off to the side is thus possible. Any little bump or any gust of wind from the side, even the famous flapping of a butterfly's wings, can cause lateral skidding (although one will probably not see any butterflies when there is ice on the road). And then the motion is one of uniform acceleration as the $t^2$ term implies. And that's exactly what you feel when you start skidding, and it is very unpleasant.

## 8.8 Some Further Remarks

Here, we'll take the opportunity to mention some parallel processes which can be either beneficial or harmful.

A layer of water on a road can create the same effect as ice. That's why traffic safety advice tells us that at a speed of 80 km/h (or 50 mph), you're waterskiing! You can even "swerve" and skid on a sandy road in the desert.

The automobile industry is currently working very hard on developing a traction control system known as anti slip regulation (ASR). Many new cars have ABS, an antilock braking system.

- ASR is currently used only when one starts to drive on slippery ice, and merely takes advantage of the fact that sliding friction is much smaller than static friction. It thus prevents the wheels from spinning when you start driving. But it would be very useful if you also had it at other times while driving so that skidding is prevented when you accelerate too quickly on ice or a wet road.

- ABS prevents the wheels from locking during braking.

The two systems need to work together in order to prevent dangerous wheel spinning or locking. That would probably prevent a lot of accidents.

Screws often begin to loosen inexplicably after a while. Maybe you didn't tighten them carefully enough. That, though, is just one possible explanation. If, for example, a machine or a cabinet door is jolted and shaken, then it's actually not the screw that is loosened but instead the surrounding material that moves, which, though, is the same thing mathematically. The movement of a screw perpendicular to the hole is virtually without any friction; the screw loosens itself.

These facts, which are so problematic for motor traffic and cabinet doors, have their advantages in other situations, though.

- We now know why you move the knife back and forth when slicing bread. You really want to cut downwards. But the lateral movement essentially cancels the friction for vertical slicing, and we reach the bottom without any problem. That's the principle of sawing.

- Anyone who has used a drill knows that the drill tends to seize up towards the end of drilling a hole. But you can pull it out again without any difficulty if you continue to rotate the drill. The motion perpendicular to the rotating motion, that is, in the outward direction, is virtually without any friction if the drill is rotating.

- Have you ever looked at yourself when you were taking a coat or a shirt off? You shake your arm because experience has shown that the item of clothing then slides down more easily. Next time, just rotate your arm left and right slightly. You'll get the same effect, and now you know why!

## 8.9 A Little Brain Teaser

$$5 + 5 + 5 = 550$$

This is clearly wrong. But you can correct it by adding a stroke.

1. Turn the $=$ into a $\neq$.
2. Turn the $=$ into a $\leq$.

There is still another possible way to get a correct equation with only the addition of one stroke.

Turn one of the plus signs into a 4 by adding a stroke on the left.

# Chapter 9

# The Snail–Racehorse Problem

## 9.1 Introduction

This chapter doesn't revolve around brave civil servants or the Big Easy. Or for that matter, any other group sent out to collect snails and for whom these little creepers turn out to be a tad too fast. No, no, quite the opposite; in this chapter, we'd like to demonstrate that these creatures are actually not that slow and can, under certain circumstances, challenge a racehorse. We need only to make one change to the location where a snail is typically found: It sits on a rubber band that is pulled by a racehorse. Well, you might say, that won't change anything. The horse will spurt ahead at a quick gallop while the snail crawls along at a breathtaking pace. They'll never come together, no matter how much they love each other. But that's wrong. Persistence permits one to reach one's goal, sometimes.

## 9.2 The Problem

Just as in the other problems, we'll need to agree on a couple of simplifications, of course.

1. Both the snail and the racehorse lead a long life, never need any food nor any sleep, and never visit the bathroom. (Sometimes the old Pony Express riders managed to ride for a whole week without a break.)

N. Herrmann, *The Beauty of Everyday Mathematics*,
DOI 10.1007/978-3-642-22104-0_9, © Springer-Verlag Berlin Heidelberg 2012

2. The rubber band can stretch by an unlimited amount. When you buy clothing, you sometimes want to have such material for your pants.

So, let's describe the problem as follows:

A perfectly elastic rubber band, which can stretch by an unlinited amount, is fixed at point 0. A snail sits there. The other end of the rubber band is tied to a racehorse at point $A$. The snail and racehorse start moving at the same time. The snail crawls at a constant speed $v_s$ along the rubber band. The racehorse flies ahead at a constant speed of $v_h \gg v_s$ and starts pulling the rubber band, which has the crawling snail on top of it.

---

**The Snail-Racehorse Problem**
(a) Can the snail catch up with the horse?
(b) How long will it take for a snail moving at $v_s = 10^{-2}$ m/s to catch up with a racehorse at $v_h = 1$ m/s, which started at point $A$ at a distance of 1 m from 0?

---

## 9.3 Mathematical Formulation

As specified in the problem, let

$v_s$  be the constant velocity of the snail,

$v_h$  be the constant velocity of the horse,

$s(t)$ be the location of the snail at time $t$.

The decisive point in all of this revolves around the fact that

speed of snail = speed of snail relative to rubber band + speed of rubber band.

You'll notice this fairly quickly if you take a rubber band and attach a paperclip to it. If you now pull on the rubber band, you'll notice that the paperclip doesn't stay at the same location but instead moves with the rubber band, even without any crawling. Every point of the rubber band therefore moves at some fraction of the horse's speed.

The speed of the snail at the location $s(t)$ is $s'(t)$, and its own velocity relative to the band is $v_s$. The speed of the point where the snail sits on the rubber band can be calculated on the basis of proportionality as follows, since we have assumed a linear elastic rubber band:

$$\text{speed of rubber band} = v_h \cdot \frac{\text{location of the snail}}{\text{location of the horse}}.$$

Because of its constant speed $v_h$ and its starting point $A$, the horse's location is given by

$$\text{location of the horse} = A + \int_0^t v_h \, dt = A + v_h \cdot t.$$

Now, we've got all the parts we need to turn the above verbal statements into a mathematical equation:

$$s'(t) = v_s + v_h \cdot \frac{s(t)}{A + v_h \cdot t}. \tag{9.1}$$

This is a linear, inhomogeneous differential equation of the first order for $s(t)$.

For readers who are not so familiar with linear differential equations (which, by the way, are sometimes referred to in abbreviated form as DEs), we'd like to offer at least a brief explanation. In (9.1), $s(t)$, i.e. the function that identifies the snail's location, is unknown. We're thus looking for a function $s(t)$ that we can substitute in the right-hand side of (9.1). The result of evaluating the right-hand side should be the derivative of the function we're looking for. This is a specific type of problem that often appears in applied mathematics. One might be tempted to believe that the entire world consists only of differential equations. Higher mathematics has developed many methods and procedures to solve such problems. Anyone familiar with this subject can skip the next section, in which we present the solution.

## 9.4 Solution of the Differential Equation

The question in this part of the task is now: "Is there a finite time $t$ at which

$$\text{location of the snail} = \text{location of the horse?}"$$

We've already mentioned the location of the horse above, so now we'll specify the location of the snail using the differential equation. To do so, we'll use a standard formula to solve the first-order linear differential equation presented in the previous section:

$$
\begin{aligned}
s(t) &= e^{-\ln\frac{A}{A+v_h t}} \cdot \int_0^t \frac{A}{A+v_h \tilde{t}} \cdot v_s \, d\tilde{t} \\
&= \frac{A+v_h t}{A} \cdot A \cdot v_s \int_0^t \frac{d\tilde{t}}{A+v_h \tilde{t}} \\
&= (A+v_h t)\left[\frac{v_s}{v_h}\ln(A+v_h \tilde{t})\right]_0^t \\
&= \frac{v_s}{v_h}(A+v_h t)\ln\frac{A+v_h t}{A}.
\end{aligned}
$$

This is the solution to the differential equation.

## 9.5 Calculating the Time of Meeting

We can now solve our main problem, i.e., that of when the horse and snail will meet, by solving the following equation for $t$:

$$0 = \text{location of the snail} - \text{location of the horse}$$
$$= s(t) - (A + v_h t)$$
$$= (A + v_h t)\left(\frac{v_s}{v_h}\ln(\frac{A + v_h t}{A}) - 1\right)$$
$$\Longleftrightarrow \ln\frac{A + v_h t}{A} = \frac{v_h}{v_s}$$
$$\Longleftrightarrow \frac{A + v_h t}{A} = e^{v_h/v_s}$$
$$\Longleftrightarrow A + v_h t = A e^{v_h/v_s}$$
$$\Longleftrightarrow t = \frac{A}{v_h}\left(e^{v_h/v_s} - 1\right).$$

If finite values of the various parameters are specified, then a finite time will be obtained for the event being that we're interested in.

> **The snail can therefore catch up with the horse.**

It's possible to clarify this surprising result by looking once again at a rubber band with a paperclip on it.

We've already seen that the paperclip doesn't stay put at one location, but instead moves with the rubber band. Every point of the rubber band thus moves at some fraction of the horse's speed.

As the snail continues to make headway, it reaches parts of the rubber band that are moved faster and faster by the horse. Eventually, the snail is literally flying with the horse because of the rubber band. And since it keeps on crawling, it ultimately reaches the horse.

## 9.6 Evaluating the Example

Let's enter the values of the parameters specified earlier into the above equation for the time $t$:

$$v = 1 \text{ m/s}, \; v_s = 10^{-2} \text{ m/s}, \; A = 1 \text{ m},$$
$$\Rightarrow \quad t = 1(e^{100} - 1) > 2^{100} = (2^{10})^{10} > (10^3)^{10} = 10^{30}.$$

This is the time in seconds. So, let's turn it quickly into years. We know that a day is equal to $86\,400$ s. If we multiply this number by

365, we get the number of seconds for a year that's not a leap year. In other words, 31 536 000. Let's make things a bit easier, though, and divide by 100 000 000, that is, by $10^8$, because a couple of seconds one way or another are not that critical.

This results in $t = 10^{22}$ years. The snail will thus have caught up with the horse after about $10^{22}$ years. Here, we need to bear in mind that the universe is approximately $10^{10}$ years old.

Let's hope that the reader is still capable of smiling after this long calculation.

## 9.7 Solution of State Capital Problem

The answer to the State Capital Problem (see p. 79) is

(see p. 79)

South Dakota,

the capital of which is

Pierre.

I hope that South Dakota will become famous throughout the world for this extraordinary property.

# Chapter 10

# The Discus Thrower Problem

An athlete throwing a discus needs to ensure that it is thrown straight ahead. The centimeters or inches which are lost when the throw is lopsided could be the athlete's downfall during a high-level competition.

## 10.1 Introduction

The beauty of mathematics is found everywhere, particularly in modern sports. We'd like to demonstrate something here with the help of a little example that might actually come as a surprise to some of you; while it's quite well known among experts, the public at large probably isn't aware of it. It's a problem that occurs when one is throwing a discus or a hammer or even doing a shot put. But if it's ignored, it could easily cost the athlete a victory.

In all of these athletic disciplines, the aim is to throw some special equipment as far as possible from a throwing circle. In so doing, certain rules need to be observed which we don't need to specify here. Our only concern is that the athletes don't have to throw their sports equipment in a very straight line. A certain amount of tolerance is permitted, which in discus throwing, for examplr, translates into an area of $20°$ to the right and left of the line straight ahead.

N. Herrmann, *The Beauty of Everyday Mathematics*,
DOI 10.1007/978-3-642-22104-0_10, © Springer-Verlag Berlin Heidelberg 2012

## 10.2 The Problem

The above fact raises the following question:

---

**The Discus Thrower Problem**
How many centimeters or inches are actually
lost when the athlete doesn't throw the discus
straight ahead, but instead a little bit to the side?

---

## 10.3 The "Loss" Formula

*Graphic(al) Derivation*

Let's look at a normal throw. If the athlete is smart enough and has
enough practice, he or she will stand directly at the point marked by
the symbol $\otimes$ in Figure 10.1 when releasing the equipment, a place
where a bar is usually sunk into the ground to act as a support for the
thrower. The distance of the throw is denoted by $w$. That's the distance
between $\otimes$ and the point where the equipment hits the ground.

But what, then, is actually measured?

We have to imagine that the measuring tape used to take the measurements runs from the point where the equipment hits the ground to
the *center M* of the throwing circle where the equipment is released.
We've already plotted this line. Now comes the crucial point. This re-
lates to the line which measures the distance of the throw from the
*circumference of the throwing circle*. In order to be able to describe
it precisely, we'll need to enlarge the sketch here – please see Figure
10.2.

Thus, the distance to be measured is different from the thrown distance (of course, we actually mean its vertical projection on the grass
surface). One can see from the sketch that a short distance is lost;
namely, the distance between the two large dots.

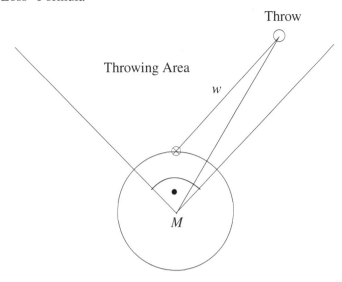

**Fig. 10.1** A normal throw. When releasing the equipment, the athlete stands at ⊗, where a sunken bar usually acts as a support for the athlete.

## *Mathematical Derivation*

In order to calculate the distance that is lost, we might recall a geometric rule, specifically the cosine law. It refers to any triangle with sides $a$, $b$, and $c$ and the corresponding angles $\alpha$, $\beta$, and $\gamma$, just as we learned in school. The cosine law establishes a relationship between the three sides and an enclosed angle. We can write it down as follows:

$$a^2 = b^2 + c^2 - 2bc \cos \alpha. \tag{10.1}$$

Now, we'll take a look at Figure 10.3.

We now apply the cosine law to the side $\widetilde{w} + r$ of the crosshatched triangle with the angle $\psi$ opposite it, which we've included in Figure 10.2. $\widetilde{w}$ is the distance of the throw, and $r$ is the radius of the throwing circle from where the equipment was released. And so, our formula reads as follows:

$$(\widetilde{w} + r)^2 = w^2 + r^2 - 2wr \cos \psi \tag{10.2}$$

We now solve the quadratic equation for the distance thrown $\widetilde{w}$:

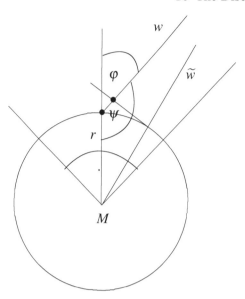

**Fig. 10.2** Enlargement of the release phase of the throw. Between the two large dots, we can see the distance that is actually lost during the throw.

$$\widetilde{w}^2 + 2\widetilde{w}r + r^2 - (w^2 + r^2 - 2wr\cos\psi) = 0.$$

We get $\psi = 180° - \varphi$, and therefore

$$\widetilde{w}_{1,2} = -r \pm \sqrt{r^2 + w^2 - 2wr\cos(180° - \varphi)}. \qquad (10.3)$$

The ominous $\pm$ has, of course, only a mathematical, not a physical, meaning here, because $-r$ is in front of it: in other words, something negative. So if we actually deduct something from it, then we'll definitely get a negative value. As a physical value, however, $\widetilde{w}$ is positive, or at least nonnegative. So we'll forget the negative sign in front of the square root in (10.3).

Thus, the distance

$$v = w - \widetilde{w} = w - (-r + \sqrt{r^2 + w^2 - 2wr\cos(180° - \varphi)}) \qquad (10.4)$$

is lost.

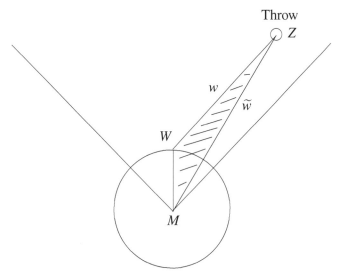

**Fig. 10.3** Here, we've added cross-hatching to the decisive triangle $\triangle MZW$. We now apply the cosine law to it.

Once again, a little test is needed to prove whether our formula is totally useless or not. Let's assume a very skillful athlete throws the equipment exactly straight ahead achieving $\varphi = 0°$. The athlete should, thus, not have lost any distance, right? If we enter this value, $\cos(180° - 0°) = \cos 180° = -1$, and we get

$$
\begin{aligned}
v &= w - (-r + \sqrt{r^2 + w^2 - 2wr\cos 180°}) \\
&= w + r - \sqrt{(r+w)^2} \\
&= w + r - (r+w) = 0.
\end{aligned}
$$

This means that the athlete didn't in fact lose anything. And our formula doesn't look too bad either.

## 10.4 Application

Let's start with discus throwing, a very old discipline which was practised by the ancient Greeks. Men have to throw a 2 kg (or 4 lb 7 oz) discus, while a women's discus weighs only 1 kg (which equals 2 lb 3 oz). The permitted area has an included angle of 40°. This means that one may deviate by 20° to the right or left of the center line.

Now, assume that an average athlete – please forgive me for having made this belittling remark, because I probably wouldn't even be able to throw the discus 20 m – has thrown the discus a distance of 50 m. In so doing, the discus drifted to the right by 20°. The throwing circle has a diameter of $2r = 2.5$ m. Thus, we get

- $r = 1.25$,
- $w = 50$,
- $\varphi = 20°$.

Now, let's turn on the calculator. By simply entering the data from our formula, we get

$$v = 50 - 49.9264 = 0.0736.$$

Consequently, the athlete will only get credit for a distance of 49.93 m, which is actually in meters; hence, our dear competitor has lost 7 cm, or 2.7559 inches. While this might cost victory at a championship, the effect can be dramatic when it comes to world records (or, rather, Earth records, since being humble might be appropriate here as well). Because that's when every centimeter or inch counts.

Now, let's calculate another example, which revolves around a female shot putter. Men have to hurl 16 British pounds, which equals 7.257 kg, while for women it's 8.8184 lb, which equals 4 kg. The throwing circle from where the shot is released has a diameter of 7 ft, which equals 2.135 m. Strangely enough, an included angle of 34.92° has been permitted since 2003 (by the way, what British measurement might that be?). At the same time, the rules and regulations for track and field events state that a 6 m wide corridor is permitted within a distance of 10 m. The author was not able to successfully establish the requisite relationship to the tangent of 34.92°/2. Never mind. Let's assume that the woman has managed the substantial distance of 20 m but, unfortunately, didn't throw the shot straight ahead. Instead, it landed right on the edge of the permitted area, at an angle of 17.46°. Our calculation is now

$$v = 20 - 19.95 = 0.05,$$

which means that 5 cm (i.e. 1.9685 inches) has been lost.

It's thus well worth the effort of investing a little time in learning how to throw the shot straight ahead.

# Chapter 11
# The Beer Coaster Problem

## 11.1 Introduction

Imagine you're sitting in a cozy pub with your friends over a nice beer. Your glasses are standing on round coasters. All of a sudden, someone takes two coasters away and holds them up. Then, for everyone to see, this person moves the two coasters on top of each other. Just a little bit at first, then more until they are exactly on top of each other, and the person thinks out loud:

"If I move the coasters on top of each other only a little, then the area covered up is very small. If I move them more and more on top of each other, then the area of overlap becomes bigger and bigger. In the end, this area is exactly the same as the area of one coaster."

## 11.2 The Problem

Now this person asks, "When is the area of overlap exactly half the area of a coaster?"

Well, that's a really interesting question! The solution is a challenge for every decent beer drinker, and can be discussed around beer tables as an essential mathematical contribution to a new world order:

N. Herrmann, *The Beauty of Everyday Mathematics,*
DOI 10.1007/978-3-642-22104-0_11, © Springer-Verlag Berlin Heidelberg 2012

> ## The Coaster Problem
> How far do you have to move two coasters on top of each
> other so that the area of the region where the two coasters
> overlap is exactly half the area of one coaster?

## 11.3 Physical Background

The way in which we find the solution here has to take account of the
usual restrictions on such tasks: when we're having a discussion in the
evening, we do'nt normally have specific resources available. Perhaps
there's somebody who has a pen and even the coasters, or maybe a
table cloth. But the formulas are often missing. So, that means we'll
have to derive all of the formulae ourselves now.

There are a few physical simplifications that we'll make to solve
this problem:

1. We look at the whole thing only in the plane; in other words, we
   consider the coasters purely as circles. That's because their thick-
   ness doesn't really matter here.
2. Next, we imagine that these infinitely thin coasters have a perfectly
   circular shape.

We need only to use a medium-sized magnifying glass in order to
recognize that these assumptions are simplifications of reality. But
that's a typical phenomenon when you want to describe nature mathematically.

## 11.4 Mathematical Description

Let's now look at the situation in the right-hand side of the sketch in
Figure 11.1.

We want the crosshatched region to the same area as each of the two
regions (half-moon-saped) that are not crosshatched. If the (common)

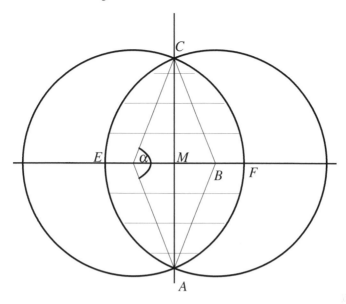

**Fig. 11.1** On the Coaster Problem

radius of the circles is $R$, then this requires the crosshatched region to have the following area:

$$A = \pi \cdot R^2 / 2. \qquad (*)$$

On the other hand, the crosshatched region consists of two (equally large) segments of circles. The area of each of these segments is equal to the difference between area of the corresponding sector with an angle $\alpha$, and the triangle $ABC$, which consists of two right triangles, and this gives us

$$A = 2 \cdot A_{\text{segment}} = 2 \cdot (A_S - A_\triangle),$$

where $A_S$ is the area of the sector and $A_\triangle$ is the area of the triangle.

We have thus introduced two unknowns at the same time, namely the radius $R$ and the angle $\alpha$. Let's hope this works! The area of the sector is calculated as follows:

$$A_S = \pi R^2 \cdot \frac{\alpha}{2\pi} = \frac{\alpha}{2} R^2.$$

The area of a right triangle is equal to half the product of the two legs. So, this results in

$$x = \overline{MB} = R \cdot \cos(\alpha/2), \; y = \overline{MC} = R \cdot \sin(\alpha/2),$$
$$A_\triangle = 2 \cdot x \cdot y/2 = x \cdot y = R^2 \cdot \sin(\alpha/2)\cos(\alpha/2).$$

Next, we insert this into our equation for $A$ and also use the equation $(*)$

$$F = 2 \cdot \left(\frac{\alpha}{2}R^2 - R^2 \cdot \sin(\alpha/2)\cos(\alpha/2)\right) = \pi \cdot R^2/2.$$

Now, let's apply some trigonometry, namely the addition theorem

$$\sin 2\gamma = 2 \cdot \sin \gamma \cos \gamma.$$

We get

$$\alpha R^2 - 2R^2 \sin \alpha = \pi R^2/2.$$

Here, it's obviously possible to cancel the factor $R^2$, which not only makes one of the two unknowns disappear, but also demonstrates that the angle $\alpha$ doesn't depend on the size of the coasters. There is a fixed universal angle $\alpha$ that causes the coasters to overlap halfway in terms of area.

**The angle $\alpha$ is a natural constant.**

And this angle can be calculated from the following equation:

$$\alpha - \sin \alpha = \pi/2. \qquad\qquad (11.1)$$

## 11.5 The Solution

Equation (11.1) is what is called a transcendental equation. It's probably impossible to solve this equation directly for the unknown angle $\alpha$. Instead, we must try to approach the solution through an approximation. Mathematics provided great procedures for this purpose a long time ago.

***The Fixed-point Procedure***

The first, fairly simple procedure that we'll discuss is called the fixed-point procedure and was introduced by Stefan Banach (1892–1945).

So, why don't you take your calculator and check out the following trick.

Adjust the calculator to "rad". Test it by entering the number 3.14159 (an approximation to $\pi$) and hitting the function key $\boxed{\cos}$. The result should be something like $-1$, maybe around $-0.999999$. Then the calculator's correct.

Now ask someone to enter any number between 0 and 1; that is a number larger than 0 but smaller than 1.

Next, ask the person to hit the $\boxed{\cos}$ key a large number of times, for instance 50 times. No matter what number the person enters, the number

$$0.7390851333$$

will always be shown on the display after a while. You could even write this number down on a piece of paper and then pretend to be a "magician." But you shouldn't repeat this; otherwise, the person will recognize that it's a trick where the number is always the same.

What's going on here? Why does it happen?

Well, Stefan Banach who introduced the whole thing in a much more general way, of course, looked at what are called fixed-point problems.

**Definition 11.1** *Given a function $T(x) : [a,b] \rightarrow \mathbb{R}$, a number $x^*$ is called a fixed point of $T$ when the following applies:*

$$T(x^*) = x^*. \tag{11.2}$$

So, this is a point that is mapped by $T$ onto itself.

We don't want to waste time with questions such as when a fixed point exists and when only one such point exists; instead, we'll just explain Banach's procedure for calculating a fixed point.

For this purpose, we can select any starting point $x_0$ from the interval $[a, b]$.

(This the first place where mathematicians have to work. They have to determine the conditions under which one can actually select any point in this interval and always get a correct result.)

$$\text{starting value:} \quad x_0.$$

Next, we calculate the first approximation with the help of

$$\text{first approximation:} \quad x_1 = T(x_0).$$

The example below will show how easy this is. And because it's so easy, we'll do it one more time:

$$\text{second approximation:} \quad x_2 = T(x_1);$$

and, once again,

$$\text{third approximation:} \quad x_3 = T(x_2);$$

and so on, again and again. We summarize the procedure in the following algorithm:

---

**Fixed-point Procedure**

1 Pick a real number $x_0$ as a starting point.

2 Calculate $x_1, x_2, \ldots$ according to the following formula:
$$x_{k+1} = f(x_k), \quad k = 0, 1, 2, \ldots$$

---

The surprising thing is this: Banach demonstrated that under certain conditions, the sequence of approximations $x_0, x_1, x_2, \ldots$ converges towards the fixed point. And the calculation is as easy as pie.

Let's look at the sketch in Figure 11.2. We have randomly selected a function $T(x)$. The bisector of the angle also shown in the sketch is the function $y = x$. In order to find the fixed point, we select a starting value $x_0$ and calculate $T(x_0)$. This value, which can be found on the

$y$ axis, becomes the new $x_1$. We've indicated this by projecting $T(x_0)$ onto the bisector of the angle and then drawing a line downward.

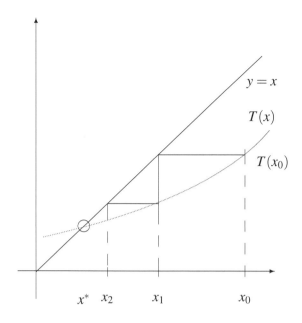

**Fig. 11.2** Sketch illustrating the fixed point procedure.

### The Newton Procedure

This procedure doesn't seek a fixed point, but instead seeks a zero of a supplied function.

Sir Isaac Newton (1643–1727) had a wonderful idea for calculating such a zero with the help of an approximation. Here comes his formula:

**Newton Procedure**

$\boxed{1}$   Select a real number $x_0$ as a starting value.

$\boxed{2}$   Calculate $x_1, x_2, \ldots$ according to the following formula:

$$x_{k+1} = x_k - \frac{f(x_k)}{f'(x_k)}.$$

The Newton procedure can be explained quite easily. Let's take a look at the sketch in Figure 11.3. We have shown the function $f$, which has a zero at $x^*$. We start at $x_0$ and go to the point $(x_0, f(x_0))$. There, we draw the tangent to the graph of $f$. This can be done quickly with the following formula:

$$\frac{y - f(x_0)}{x - x_0} = m = f'(x_0) \implies y - f(x_0) = (x - x_0) \cdot f'(x_0).$$

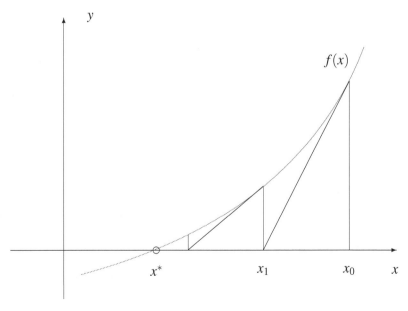

**Fig. 11.3** Sketch on the Newton Procedure

We now find the intersection of this tangent with the $x$ axis; in other words, we equate $y$ to 0. We then use this intersection as the new approximation $x_1$. The requisite formula is as follows:

$$x_1 = x_0 - \frac{f(x_0)}{f'(x_0)}.$$

But that's exactly the above Newton formula.

## 11.6 Application to the Beer Coaster Problem

Now, we can try to evaluate the solution of our beer coaster problem.

First, a rough sketch (Figure 11.4) will help us to get an idea of the solution. We need to change (11.1) a little:

$$\alpha - \pi/2 = \sin\alpha.$$

We sketched both sides of this equation; that is, the functions $y(\alpha) = \alpha - \pi/2$ and $y(\alpha) = \sin\alpha$. W can see that there's just one solution to our beer coaster problem.

### *Solution by the Fixed-point Procedure*

Let's try first with a fixed-point calculation. In order to do so, we need to put the problem into fixed-point form. A simple idea is to transform (11.1) as follows:

$$\alpha - \sin\alpha = \pi/2$$
$$\Longleftrightarrow \quad \alpha = \sin\alpha + \pi/2. \qquad (11.3)$$

This is of the form $x = T(x)$, but with $\alpha$ in place of $x$.

We take a rough starting value $\alpha_0 = 2$ from the sketch. When we calculate the first approximation value, we get

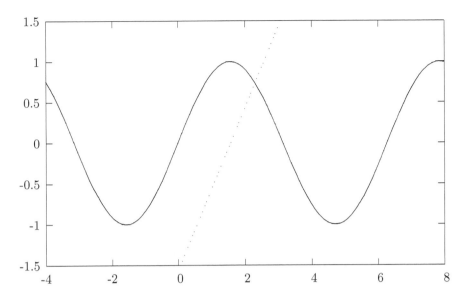

**Fig. 11.4** Sketch for the beer coaster equation (11.1). As you can see, the graphs of $y(\alpha) = \alpha - \pi/2$ and $y(\alpha) = \sin(\alpha)$ intersect exactly once, near $\alpha = 2$.

$$\alpha_1 = \sin \alpha_0 + \pi/2 = 2.5.$$

We then get

$$\alpha_2 = \sin \alpha_1 + \pi/2 = 2.18509.$$

And the game continues in the same way. If we put our little calculator to work again, we end up with

$$\alpha_{35} = \sin \alpha_{34} + \pi/2 = 2.309881.$$

If we continue to iterate, it doesn't change the result. So we'll take this as the final result.

*Solution by the Newton Procedure*

Now, we calculate the zero point of the following equation by the Newton procedure:

$$\alpha - \sin \alpha - \pi/2 = 0.$$

We write

$$f(\alpha) = \alpha - \sin \alpha - \pi/2,$$

and thus

$$f'(\alpha) = 1 - \cos \alpha.$$

We start with $\alpha_0 = 2$ again. This gives us the following result for $\alpha_1$:

$$\begin{aligned} \alpha_1 &= \alpha_0 - \frac{f(\alpha)}{f'(\alpha)} \\ &= \frac{\alpha_0 - \sin \alpha_0 - \pi/2}{1 - \cos \alpha_0} \\ &= 2.339013. \end{aligned}$$

Once again, we use our little calculator to do the additional calculations. The results are:

$$\begin{aligned} x_0 &= 2.000000 \\ x_1 &= \underline{2.3}10063, \\ x_2 &= \underline{2.309881}, \\ x_3 &= \underline{2.309881}, \\ x_4 &= \underline{2.309881}. \end{aligned}$$

Wow, that's much faster than the fixed-point method! But it would have been a surprise if that were not the case, because we have to solve a much more complicated equation, including a derivative. Note that in the above, we've underlined the digits that won't change anymore.

In both cases, we thus get the following final value for the angle in radians:

$$\alpha = 2.309881.$$

## 11.7 Concluding Remarks

If we convert the final value obtained from the above iteration from radians to degrees, then we get

$$\alpha = 132.3464473° \approx 90° + 42°.$$

Wasn't there something about the number 42?? Yes, there was a certain Mr. Adams (see [1]) who selected 42 as the ultimate number of the universe. That seems just to have been a fabrication; in any case, no reason was given why that number had been selected. But with this beer coaster story, we've supplied a reason why the "meaning of life" can be found in a simple party problem.

We should also mention that the length of the line $l = \overline{EF}$ on the $x$ axis, which might be used as an indicator of the degree of overlap, depends on the radius of the beer coasters, of course. For this, we have the equation

$$l = 2R - 2\overline{MB} = 2\left(R - R\cos\frac{\alpha}{2}\right).$$

# Chapter 12
# The Toasting Problem

## 12.1 Introduction

Once again, another awful New Year's Eve party where one tends to make so many New Year's resolutions around midnight, while at the same time, the resolutions of the old year haven't yet been accomplished. So, you sit and worry about how to improve the situation when you're about to toast all the others and the ball is about to drop at Times Square.

And then you get this grand idea that will make the party take off. Let's just ask a simple question:

Thirty people have gathered here. And, in a minute, everyone is to toast one another. How often will the glasses clink? This will probably result in a few frowns. But it sounds so easy. Well, probably 30 times. But everyone with everyone else, is that maybe 60 times? But one doesn't really toast oneself, so could it maybe be only 59 times?

Well, a mathematician can solve this right away.

N. Herrmann, *The Beauty of Everyday Mathematics*,
DOI 10.1007/978-3-642-22104-0_12, © Springer-Verlag Berlin Heidelberg 2012

## 12.2 The Problem

---

**The Toasting Problem**

If $N$ people stand together and everyone toasts
everyone else, how often will the glasses clink?

---

Let's tackle this problem logically.

1. If only one person is in the room, then that person doesn't toast anyone but instead drowns their own sorrows.
2. If two people are present, then the glasses clink once. Whatever other tension might also be present we'll just ignore, because we just want to do math.
3. With three people, let's think this way. The two persons are joined by a third person. This person has to toast the other two; in other words, two more times plus the one toast of the first two people. This equals

$$1 + 2 = 3.$$

4. Now, we're courageous and look at four people. Let's assume that three people are already there, for whom we have already counted three clinks. The fourth now enters and must toast the other three, which gives us

$$1 + 2 + 3 = 6.$$

5. Now, it becomes a bit clearer. With five people, it's

$$1 + 2 + 3 + 4 = 10$$

clinks.

But here comes the mathematician:

How often do the glasses clink for $N$ people?

Well, does this ring a bell? Correct, for $N$ people, it's

$$1+2+3+\cdots+(N-1).$$

In other words, you have only to add the numbers together. Now comes the tricky part for mathematicians. Is there maybe a simple formula for this operation?

Let's try to approach this topic with an example. What do you think of the following question:

How big is the sum of the first 100 numbers, that is, $1+2+\cdots+$ 100?

That's the famous question that a teacher once asked little Carl Friedrich[1] in order to keep him busy for a while. Back then, teachers often had to teach a number of different classes at the same time. This was accomplished by giving some of the students longer assignments, which they had to solve by themselves while the teacher continued to work with the other students. But the teacher didn't take Carl Friedrich's abilities into account. He solved the problem in the blink of an eye ("The sum is 5050!") and continued to annoy the teacher.

How did he do it? Well, maybe the whole thing was just an anecdote and never really happened. But we can find an easy way to add the one hundred numbers in our head. Whether Gauß did it the same way back then isn't known today.

We divide the one hundred numbers into four blocks. The first block goes from 1 to 49, the second contains only 50, the third goes from 51 to 99, and the fourth contains only 100. We write the first and the third block next to one another, but write the third block from the top down:

$$
\begin{array}{cc}
1 & 99 \\
2 & 98 \\
\vdots & \vdots \\
48 & 52 \\
49 & 51 \\
\end{array}
$$

We see immediately that every line adds up to 100. That's 49 times 100. Then, there is the 100 from the fourth block. So, that's 50 times 100, which equals 5000. So, now, there's only the 50 from the second

---
[1] Gauß, C.F. (1777–1855).

block floating around somewhere. If we quickly add it, we get 5050 for the sum of the numbers between 1 and 100. Simply ingenious!

Can we generalize this?

We divided the one hundred numbers into four blocks by, essentially, making a break halfway. Let's assume that the final number is even. In the case above, the final number was 100, and the halfway point was 50. We then saw that the sum equaled 50 times 100 by placing the blocks next to one another. In other words,

$$\frac{n}{2} \cdot n.$$

We still have to add our lonely middle element $n/2$, which results in the formula

$$\Sigma = \frac{n}{2} \cdot n + \frac{n}{2}.$$

That looks like a neat formula, which we can actually write much more simply:

$$\Sigma = \frac{n}{2} \cdot n + \frac{n}{2} = \frac{n(n+1)}{2},$$

which you can probably follow quite easily. We've come up with this formula for the case when $n$ is an even number. That makes it possible to find the number in the middle. But what about an odd $n$?

We'll just simply assume that our formula is always right, even for odd numbers. In other words, we have the following theorem.

**Theorem 12.1** *For the sum* $1 + 2 + \cdots + n$, *the result is always*

$$\Sigma = 1 + 2 + 3 + \cdots + n = \frac{n \cdot (n+1)}{2} \quad \text{for } n \geq 1.$$

That's quite a big claim. After all, it's supposed to be correct for all natural numbers; in other words, for an infinite set of numbers. And that's a really big number! But Mathematicians have come up with a great way of proving such formulas.

## 12.3 Mathematical Induction

Let's begin with an example that's not meant to be taken too seriously.

**Assumption:** *60 is divisible by all numbers.*

Let's try it like a physicist:

60 is divisible by 2, ok.
60 is divisible by 3, also ok.
60 is divisible by 4, by 5, by 6.
60 is divisible by 7? No, but that's probably a measurement error.
Let's take one more sample just to be sure:
60 is divisible by 10.
Everything's fine. Except for one measurement error, there's no contradiction.

Yes, that's a typical mathematician's gibe!

In te case of the following assumption, you've got a little more work to do.

**Assumption:** *The function*

$$f(x) = x^2 - x + 41$$

*provides a prime number for every natural number x.*

Some examples show the following:

- $f(3) = 47$ is a prime number.
- $f(10) = 131$ is a prime number.
- $f(20) = 421$ is also a prime number.

We'll reveal the answer to you. You have to have a lot of patience. Until the number 40, there'll always be prime numbers. It's only the number 41 which destroys the assumption:

$$f(41) = 41 \cdot 41 - 41 + 41 = 41 \cdot 41.$$

That's not the right way. It's not actually a problem to come up with questions where you have to work for a long time in order to detect a contradiction. Calculating individual examples can sometimes be amusing, but it doesn't prove anything. It only works if you're looking for the opposite of the assumption you're examining.

We need a comprehensive principle to prove statements for all natural numbers. Calculating individual examples doesn't really prove anything. Mathematical induction can help here. The principle of proof by mathematical induction might sound, when first encountered, to be very doubtful. A statement might read as follows.

---

**Principle of Mathematical Induction**

*If a statement is correct for $k = 1$ and if this correctness for a natural number $k$ (a prerequisite for induction) always results in correctness for the subsequent number $k + 1$, then the statement is correct for every natural number.*

---

That's not actually so difficult to understand. If I know something for $k = 1$ and if I can always prove that it applies to the next number, then it also applies to 2 and then to 3 and then to 4, etc. In other words, it applies to all natural numbers.

We therefore have to consider two aspects:

1. We have to find an initial case for which our assumption is correct: the base case of the induction.
2. We have to draw general conclusion that allows us to argue from $k$ to $k + 1$: the inductive step.

Our above examples demonstrate that we can't avoid the inductive step.

Take a look at the following assumption: All people have the same hair color, namely mine. Please excuse my selfishness, but maybe you could send me a sample of your hair, and then I can use it as a prime example in the next edition.

For the base case we'll take a set of people that consists only of me. Then, all people in this set have my hair color.

Let's suppose that the assumption is correct for $k$ people. Then we take a set of $k+1$ people. Without any great restrictions, we can assume that I'm among these people. Then, we assume that I'm thrown out of this set and $k$ people remain, who, according to the assumption, all have my hair color. So the $k+1$ people, once I'm allowed back into the circle, all have my hair color as well. And a "proof" has been presented.

Our mistake is to be found in the base case. We actually want to compare all people. So it's not enough to focus just on one, i.e., me. As soon as we have a set of two people, of whom I'm one, it becomes difficult to demonstrate that the other person has my white hair. So the base case was chosen mistakenly.

Let's return to our addition, for which we have asserted that

$$\Sigma = 1+2+3+\cdots+n = \frac{n\cdot(n+1)}{2} \quad \text{for } n \geq 1.$$

For $k=1$, we only have 1 in the sum. The right-hand side of the formula equals $\frac{1(1+1)}{2} = 1$, which is correct.

Let's do this also for $k=2$. On the left-hand side, we get $1+2=3$. If we check the right-hand side, we find that: $\frac{2(2+1)}{2} = 3$. So here it looks quite reasonable as well.

Now comes the difficult part.

We assume that the formula is correct for any $k$. Then we have to prove its correctness also for $k+1$. In other words, we have to show that

$$\underbrace{1+2+\cdots+k}+(k+1) = \frac{(k+1)(k+2)}{2}.$$

In so doing, we can and must take advantage of the fact that we know the sum on the left-hand side with the brace underneath it. With this knowledge, we continue to calculate:

$$1+2+\cdots+k+(k+1) = \frac{k(k+1)}{2}+k+1$$
$$= \frac{k(k+1)}{2}+\frac{2(k+1)}{2}$$
$$= \frac{k(k+1)+2(k+1)}{2}$$
$$= \frac{(k+1)(k+2)}{2},$$

and that's exactly what we wanted to prove.

With this trick, we can now expand our knowledge from $k=1$ to $k=2$, then to $k=3$, then to $k=4$, etc. In other words, we know the result for all natural numbers.

## 12.4 Application

Now, let's return to the toasting problem. There's just one small thing to consider, though. Someone alone can't toast anyone, since no one else is there. In other words, our task only makes sense if we have at least two people. These two people toast exactly once. With $N$ people, we therefore have to consider that the $N$th person toasts $N-1$ persons. This means that our sum goes only as far as $N-1$. Mathematically speaking, our toasting problem is as follows:

How big is the sum

$$1+2+\cdots+(N-1)?$$

According to our formula, this sum is

$$1+2+\cdots+(N-1) = \frac{(N-1)\cdot N}{2}.$$

Here are some examples. With 30 people, the glasses clinky

$$\frac{29\cdot 30}{2}=435$$

times That's quite a lot of clinking.

With 50 people, the number rises to $\frac{49 \cdot 50}{2} = 1225$ times.

## 12.5 Related Problems

This problem of evaluating the sum of $n$ numbers also appears elsewhere.

For example, one can ask how often $N$ pairwise different straight lines in $\mathbb{R}^3$ can intersect one another. One can't exclude the possibility that all of the lines run parallel to one another; then they don't intersect at all. In other words, we already know the minimum number. But we can rule out the possibility that two straight lines coincide with one another, because they have to be different pairwise. Otherwise, we'd have an infinite number of common points. So, what then is the maximum number of intersection points when the lines are pairwise different?

The answer can be copied literally from above. "Intersecting each other" corresponds exactly "toasting with one another". We thus get the same answer as the one stated above in Theorem 12.1: $N$ pairwise different straight lines in $\mathbb{R}^3$ intersect each other at

$$\frac{(N-1) \cdot N}{2}$$

points at most.

For another related problem, consider Figure 12.1.

Here, we have half a square, with sides of length 10 units. We've drawn a lot of little squares inside it, which all have the same size of one unit. Now, we ask how many such squares there are, for a side length of arbitrary $N$, of course.

In our example with a side length of $N = 10$, we have one square in the top row, below it, two squares, etc., until we can count nine squares in the bottom row. Together, that's $1 + 2 + \cdots + 9 = \frac{9 \cdot 10}{2} = 45$ squares according to our formula.

With our little game of squares, we can actually derive the formula for the sum in a totally different way. For a side length $N$, the full square would contain $N \cdot N$ squares. The diagonal, however, doesn't

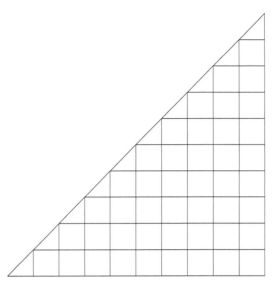

**Fig. 12.1** Game of squares to calculate $1 + 2 + \cdots + (N-1)$.

cooperate. It contains $N$ squares. We have to allocate the difference $N \cdot N - N$ to the top and the bottom halves, in other words, divide by 2. We get:

$$\Sigma = \frac{N \cdot N - N}{2} = \frac{(N-1) \cdot N}{2},$$

which is exactly our old formula.

# Chapter 13
# The Heart Problem

## 13.1 Introduction

This last chapter is dedicated to heartache. Each and every one would like to give away their heart at least once in their lifetime. What creative ideas might a mathematically interested person have? At this point, one needs to activate one's brain, because you shouldn't fool around with such questions.

## 13.2 The Problem

> **The Heart Problem**
> How does a mathematician declare his or her
> love in a proper and decent manner?

## 13.3 First Solution

It won't come as a complete surprise that the answer is a formula:

"Oh, darling, I'll give you my

$$y = |x| \pm \sqrt{1 - x^2}, \quad -1 \le x \le 1."$$  (13.1)

N. Herrmann, *The Beauty of Everyday Mathematics*,
DOI 10.1007/978-3-642-22104-0_13, © Springer-Verlag Berlin Heidelberg 2012

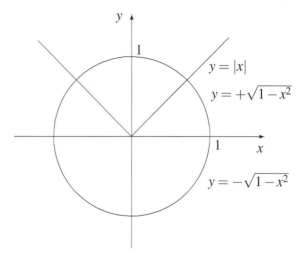

**Fig. 13.1** Creating the first cardioid. We have drawn graphs of the two func-
tion $y = |x|$, which is two half-rays bisecting the upper two quadrants, and
$y = \pm\sqrt{1-x^2}$, which is the unit circle.

You need to take a closer look to recognize the relatively simple
concept behind this formula. First of all, let's keep the root as simple as possible in this formula. So, the expression under the root sign
shouldn't be negative. This is why we restrict $x$ with the condition
$-1 \leq x \leq 1$.

Then, there are two components, which we'll interpret as two functions

$$y = |x|$$

and

$$y = \pm\sqrt{1-x^2}.$$

We should recognize the first one from school. For $x \geq 0$, i.e., in the
right half-plane, it is the bisector of the first quadrant, and for $x < 0$,
in the left half-plane, it is the bisector of the second quadrant. These
are both above the $x$ axis.

The second function is also known, but we need to change the equation for it a little by squaring it:

$$y = \pm\sqrt{1-x^2} \Longrightarrow x^2 + y^2 = 1.$$

This is the equation for the unit circle; the upper half of the circle is obtained with the plus sign and the lower half with the minus sign.

The graphs of the two functions are drawn on the same system of coordinates in Figure 13.2.

We now have to add the graphs of the two functions together. To the graph of $y = |x|$, we add first the upper half of the circle, and then to the bottom half. The result is shown in Figure 13.2. Once again, everything is drawn on one system of coordinates.

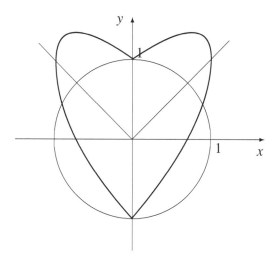

**Fig. 13.2** The First cardioid, according to the formula $y = |x| \pm \sqrt{1-x^2}$, $-1 \le x \le 1$

Isn't that a heartwarming heart? Your little daughter will love you if you put it in her friendship book. But be careful. The author speaks from experience. All of her girlfriends will then also want to have something like this in their friendship books. This could create logistical problems. First of all, you'll need a lot of time to organize everything, and then you'll need many ideas. But then again, maybe this book will help you.

We've drawn everything one more time in Figure 13.3 to show the end result without mathematical details. We've even used little diamonds instead of points in order to create a necklace in the shape of a little heart.

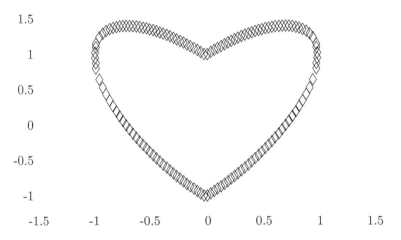

**Fig. 13.3** Our cardioid according to the formula $y = |x| \pm \sqrt{1-x^2}, \; -1 \le x \le 1$, changed a bit with little diamonds to create the illusion of a necklace

## 13.4 Additional Solutions

Mathematicians are creative people. You can find many other beautiful depictions of hearts on the Internet.

For example, http://www.schulmodell.de/mathe/herz includes the following formula:

$$y = \frac{2}{3} \cdot \left( \frac{x^2 + |x| - 6}{x^2 + |x| + 2} \pm \sqrt{36 - x^2} \right)$$

Analogously to our previous formula, we again have to restrict the variable $x$ so that the root behaves properly (Figure 13.4) and doesn't complicate matters:

$$-6 \leq x \leq 6.$$

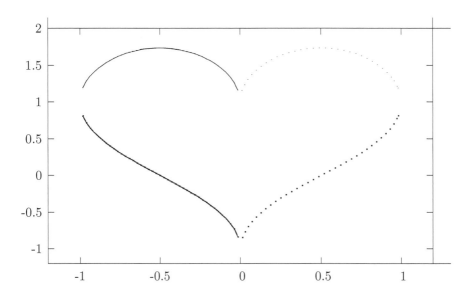

**Fig. 13.4** A cardioid found on the Internet according to the rather complex formula $y = \frac{2}{3} \cdot \left( \frac{x^2+|x|-6}{x^2+|x|+2} \pm \sqrt{36-x^2} \right)$.

Those of you who'd like a depiction of a heart in three dimensions should try the following formula:

$$(2x^2 + y^2 + z^2 - 1)^3 - \frac{x^2 \cdot z^3}{10} - y^2 \cdot z^3 = 0.$$

The plotting might be easier if we give you the solution of this equation for $z$:

$$z_{1,2} = \frac{\sqrt[3]{\frac{x^2}{10}+y^2}}{2} \pm \sqrt{\frac{\left(\sqrt[3]{\frac{x^2}{10}+y^2}\right)^2}{4} - 2x^2 - y^2 + 1}.$$

The best way to illustrate this would be a 3-D plot in color. But that would go well beyond the scope of this little book.

# References

1. Adams, D.: *The Hitchhiker's Guide to the Galaxy*, Serious Productions Ltd., 1979.

2. Adams, D.: *Per Anhalter durch die Galaxis*, Heyne, Munich, 1998.

3. Herrmann, N.: *Höhere Mathematik für Ingenieure*, Aufgabensammlung, Bd. I, II, Oldenbourg, Munich, 1995.

4. Herrmann, N.: *Höhere Mathematik für Ingenieure, Physiker und Mathematiker*, Oldenbourg, Munich, 2004.

5. Herrmann, N.: *Können Hunde rechnen?*, Oldenbourg, Munich, 2007.

6. Luus, R.: How to remember $\pi$?, University of Toronto, personal Communication, 1985.

7. Meschkowski, H.: *Schüler-Duden-Mathematik* I, II, Bibliogr. Institut Mannheim, 1972.

8. Meschkowski, H.: *Aufgaben zur modernen Schulmathematik mit Lösungen*, Bibliogr. Institut Mannheim, 1970.

9. Vogel, H.; Gerthsen, C.: *Physik*, Springer, Berlin, Heidelberg, 1995.

10. Walker, J.: *Der fliegende Zirkus der Physik*, Oldenbourg, Munich, 2000.

11. Wille, F.: *Humor in der Mathematik*, Vandenhoek & Ruprecht, Göttingen, 1984.

12. Wode, D.: Wie ein Auto bei Glätte rutscht, Math. Sem. Ber. **35**, 1988.

N. Herrmann, *The Beauty of Everyday Mathematics*,
DOI 10.1007/978-3-642-22104-0, © Springer-Verlag Berlin Heidelberg 2012

# Index